Introduction to Renewable Energy for Engineers

Introduction to Renewable Energy for Engineers

KIRK D. HAGEN
Weber State University

PEARSON

Hoboken • Boston • Columbus • San Francisco • New York
Indianapolis • London • Toronto • Sydney • Singapore • Tokyo • Montreal
Dubai • Madrid • Hong Kong • Mexico City • Munich • Paris • Amsterdam • Cape Town

Vice President and Editorial Director, ECS: *Marcia J. Horton*
Executive Editor: *Holly Stark*
Editorial Assistant: *Michelle Bayman*
Program and Project Management Team Lead: *Scott Disanno*
Program Manager: *Erin Ault*
Project Manager: *Sandra L. Rodriguez*
Global HE Director of Vendor Sourcing and Procurement: *Diane Hynes*
Director of Operations: *Nick Sklitsis*
Operations Specialist: *Maura Zaldivar-Garcia*
Product Marketing Manager: *Bram Van Kempen*
Field Marketing Manager: *Demetrius Hall*
Marketing Assistant: *Jon Bryant*
Cover Designer: *Black Horse Designs*
Manager, Rights and Permissions: *Rachel Youdelman*
Full-Service Project Management: *Saraswathi Muralidhar, Lumina Datamatics, Inc.*
Composition: *Lumina Datamatics, Inc.*
Typeface: *10/12 New Baskerville Std-Roman*

Pearson Education Ltd., London
Pearson Education Singapore, Pte. Ltd
Pearson Education Canada, Inc.
Pearson Education—Japan
Pearson Education Australia PTY, Limited
Pearson Education North Asia, Ltd., Hong Kong
Pearson Educación de Mexico, S.A. de C.V.
Pearson Education Malaysia, Pte. Ltd.
Pearson Education, Inc., Hoboken, New Jersey

Library of Congress Cataloging–in–Publication Data
Hagen, Kirk D.

   Introduction to renewable energy for engineers / Kirk D. Hagen, Weber State University.
      pages cm
   Includes bibliographical references and index.
   ISBN 978-0-13-336086-8 – ISBN 0-13-336086-5
  1. Renewable energy sources.   2. Sustainable engineering.   I. Title.
   TJ808.H34 2015
   621.042—dc23
                              2015015702

PEARSON

ISBN 10: 0-13-336086-5
ISBN 13: 978-0-13-336086-8

Contents

6 • MARINE ENERGY   121

7 • BIOMASS   136

APPENDIX A • THE TRAPEZOIDAL RULE   150

APPENDIX B • ANSWERS TO SELECTED PROBLEMS   153

INDEX   155

About This Book

Introduction to Renewable Energy for Engineers is intended for beginning engineering students and students in other fields of study who want to learn the fundamental engineering principles of renewable energy. Following an introductory chapter that briefly covers the main types of renewable energy, the basics of energy and power calculations, and the fundamental economics of renewable energy systems, the book devotes a separate chapter to each renewable energy type: solar, wind, hydro, geothermal, marine, and biomass. The primary focus of this book is the application of renewable energy to electrical power generation. As each renewable energy technology is explained, the student is shown how to do a basic energy analysis of the corresponding power-generation system. Each chapter concludes with a discussion of the environmental impacts of the renewable energy type covered in that chapter. Finally, a set of problems is given at the end of each chapter for the student to work to reinforce the concepts taught in the chapter.

ABOUT THE AUTHOR

Kirk D. Hagen is a professor of engineering at Weber State University in Ogden, Utah, where he has taught since 1993. He did his undergraduate work in physics and earned his PhD in mechanical engineering at the University of Utah. His areas of specialization are thermodynamics, fluid mechanics, and heat transfer. *Introduction to Renewable Energy for Engineers* is his fourth book. He is also the author of *Heat Transfer with Applications, Heat Transfer Solutions: Worked Problems to Supplement a First Course in Engineering Heat Transfer,* and *Introduction to Engineering Analysis.*

1 Introduction to Renewable Energy

Objectives

After reading this chapter, you will have learned:

- What renewable energy is
- How solar energy systems work
- How wind energy systems work
- How hydroenergy systems work
- How geothermal energy systems work
- How marine energy systems work
- How biomass energy systems work
- How to do basic energy and power calculations
- How to do basic economics calculations
- The basic environmental issues with renewable energy

1.1 INTRODUCTION

Energy has always been crucial to the survival and development of humankind. Early humans used the energy of their own bodies to walk and run, hunt, build shelters, and transport their belongings, and they used the energy of fire to keep warm and cook food. As the earth's population grew and societies developed, the energy of humans was augmented by the energy of animals that were tamed for livestock. The transition from a hunting to an agricultural society necessitated the use of livestock for planting crops and transporting water and supplies. As cultures became more sophisticated, humans learned how to use fire to smelt ore for making copper, iron, and other metals. The energy produced by wind and naturally flowing water was harnessed to grind grain for food, pump water for irrigation, and saw timbers for construction. The steam engine, a development of the industrial revolution, produced mechanical energy for a variety of uses, including transportation, water pumping, and manufacturing. In the modern world, energy is used to operate electrical devices, power transportation systems, and heat and cool living spaces.

Before we define renewable energy, let us define some key technical terms—*energy* and *power*. A concise definition of **energy** is the *capacity to do work*. If a device or system has the capacity to do work, it possesses at least one form of energy that is available for transformation to another form of energy. A simple illustration is a stretched spring. When a stretched spring is released, its stored (potential) energy is transformed into energy of motion (kinetic energy). If, for example, the spring was connected to a component in a mechanism, work would be done on that component. Another illustration of a system in which an energy transformation occurs is a wind turbine. A wind turbine is basically an electrical generator with a set of blades mounted on its shaft. Wind causes the generator's shaft (the armature of the generator) to rotate. As the wire windings of the armature cut the magnetic field lines of magnets that surround the armature, a voltage is produced. Hence, the kinetic energy of the wind is transformed into electrical energy, which is a form of work. The unit of energy in the SI and English unit systems is the joule (J) and British thermal unit (Btu), respectively. **Power** is the *time rate of energy*. Returning to the stretched spring example, suppose that as the stretched spring returns to its original length, it produces an amount of kinetic energy during a certain period of time. This quantity is the power of the spring and is obtained by dividing energy by time. The unit of power in the SI and English unit systems is the watt (W) and horsepower (hp), respectively. Energy and power calculations are discussed in more detail at the end of this chapter.

Renewable energy is *energy that comes from sources that are naturally replenished*. Stated another way, renewable energy comes from sources that are refilled with energy at rates comparable to the rates of energy extraction. Energy from sources that, for all practical purposes, are inexhaustible is also considered renewable. Solar energy is one form of renewable energy. The amount of solar energy intercepted by the earth each year is approximately 5.4×10^{24} J. Therefore, the amount of solar energy that is absorbed by the earth and reradiated plus the solar energy that is reflected from the earth each year is approximately 5.4×10^{24} J. If the energies entering and leaving the earth were not equivalent, the temperature of the earth's surface would increase or decrease, depending on the relative amounts of the incoming and outgoing energies. The average temperature of the earth's surface is fairly stable at about 14°C (57°F). Wind energy is also a form of renewable energy. Winds are caused by pressure differences in the atmosphere induced primarily by uneven heating of the earth's surface by the sun. Hence, wind energy is a form of solar energy. Other forms of renewable energy are hydro, geothermal, marine, and biomass. Each renewable energy form is introduced in the next section.

Renewable energy is a field that is subsumed by **sustainability**, a broad interdisciplinary field whose objective is to develop and maintain a diverse and productive ecosystem. Sustainability, sometimes referred to as *sustainable development*, encompasses and connects environmental, societal, and economic issues in an attempt to meet the current needs of humankind without compromising the needs of future generations.

Energy, most notably renewable energy, is a critical component of sustainability. Climate scientists of the Intergovernmental Panel on Climate Change (IPCC) recently concluded that the probability is at least 90 percent that the atmospheric increase in global carbon dioxide, a greenhouse gas that induces climate change, is human caused. The IPCC states that the primary cause of the atmospheric CO_2 increase is fossil-fuel emissions. Reducing these emissions is a multidimensional problem that can be addressed on commercial and personal levels. Unlike power plants that burn fossil fuels, power plants that operate on renewable energy technologies do not release CO_2 into the atmosphere, so the implementation of renewable energy technologies is an important part of the solution to this problem.

The primary advantages of renewable energy sources are that they are sustainable, found nearly everywhere across the globe, and are essentially nonpolluting. Unlike power plants that burn fossil fuels, some renewable energy-based power plants, such as photovoltaic solar panels and wind turbines, do not require water for their operation. Some disadvantages of renewable energy sources include energy variability and low energy density. For example, wind speeds vary widely in many locations, and prevailing weather conditions prevent some regions from receiving significant amounts of solar energy. Hence, to achieve comparable power productions of conventional energy sources, higher initial costs are often encountered with renewable energy systems.

Fossil fuels—coal, oil, and natural gas—are not renewable energy sources because these materials take an extremely long time to be replenished in the earth. Nuclear fuels are not renewable energy sources because these materials are finite in the earth and cannot be replaced. According to the International Energy Agency (IEA), over 31 Gt (gigatons) of CO_2 were emitted globally from fossil-fuel combustion in 2011. The IEA also points out that although coal represented 29 percent of global energy supply in 2011, it accounted for 44 percent of the global CO_2 emissions because of its heavy carbon content per unit of energy released. A typical coal-fired power plant is shown in Figure1.1.

Figure 1.1
Coal-fired power plants produce more electrical power in the United States than any other single type of power plant.

(Martin33/Fotolia)

Electrical energy is the primary form of energy used by societies, and there are three types of power plants or sources from which electrical energy is derived—conventional thermal, nuclear, and renewables. According to the U.S. Energy Information Administration (EIA), approximately 66 percent of the world's electricity in 2010 was generated by conventional thermal power plants, about 8 percent by nuclear power plants, and around 26 percent by renewables. A conventional thermal power plant converts thermal energy to electrical energy by burning fossil fuels. A nuclear power plant accomplishes the same task, but the thermal energy originates from fission processes. The reason that the percentage of electricity generated by renewables is relatively large is because hydropower is a major source of electrical power. Approximately 78 percent of global electricity from renewables in 2010 was generated from hydropower, about 14 percent from wind, about 5 percent from biomass, approximately 2 percent from solar and marine, and about 1 percent from geothermal.

The sources of renewable energy covered in this book are solar, wind, hydro, geothermal, marine, and biomass. In the sections that follow, brief descriptions of these renewable energy sources are given.

1.2 SOLAR ENERGY

Solar energy emanates from the sun, the yellow star at the center of our solar system. Like all stars, the sun generates vast amounts of energy by nuclear fusion reactions in its interior. The energy from these reactions is transported to the sun's surface where it radiates into space. The sun has been radiating energy at a reasonably steady rate for several billion years and is predicted to continue to do so for billions of years to come, which is the basis for considering solar energy as a renewable source. The sun radiates energy at a rate of 3.9×10^{26} W. At the outer edge of the earth's atmosphere, the average solar heat flux (solar power intercepted by a plane surface facing directly into the sun) during the year is approximately 1366 W/m^2. This quantity, referred to as the *solar constant*, is based on recent solar heat flux measurements and is subject to revision as new measurements are made. Because of absorption by atmospheric carbon dioxide and water vapor and interactions with dust and other pollutants, the power density represented by the solar constant does not reach the earth's surface. Thus, the power input for all earth-based solar energy systems is less than the solar constant.

A portion of the solar energy that reaches the earth's surface can be collected and used to heat water or a living space or to generate electrical energy. This collection process involves either a solar panel that concentrates the sun's rays on a fluid, thereby heating the fluid to run a turbine generator, or a solar panel that converts solar energy directly into electrical energy. These two types of solar energy systems are described in Chapter 2. A solar power plant of the second type is shown in Figure 1.2.

1.3 WIND ENERGY

Wind is the movement of atmospheric air across the earth's surface from regions of high pressure to regions of low pressure. The primary cause of winds is the uneven heating of the earth's surface by the sun, which depends on latitude, time of day, and the distribution of land and large bodies of water, particularly the oceans. Another cause of winds is fluid friction between the atmosphere and earth's surface, which allows the earth to drag the atmosphere around producing

turbulence. Horizontal components of wind velocities are normally much greater than the vertical velocity components. The kinetic energy of the wind, and therefore the wind's power-generating potential, is proportional to the cube of wind velocity. Because winds are primarily caused by uneven heating effects of the sun, wind energy is considered to be an indirect form of solar energy and is therefore renewable.

Kinetic energy of the wind is converted to electrical energy using a wind turbine. There are primarily two types of wind turbines, each being characterized by the orientation of the axis or shaft. A horizontal axis wind turbine (HAWT) typically consists of a set of three blades mounted to a horizontal shaft that is connected to an electrical generator. This traditional "windmill"-style turbine is used in a variety of applications from 5-MW wind farms to 100-kW residential applications. A vertical axis wind turbine (VAWT) resembles an "eggbeater" and typically consists of three blades mounted to a vertical shaft. VAWTs are primarily used in small-scale applications and are less common than HAWTs. These two types of wind turbines are described in Chapter 3. A wind farm consisting of HAWTs is shown in Figure 1.3.

1.4 HYDROENERGY

Hydroenergy refers to energy derived from flowing or falling water. The terms *hydropower* and *hydroelectric* are commonly used in conjunction with this type of renewable energy source. The kinetic energy of water flowing in a river or stream can be harnessed directly to produce work. More commonly, the flow of water is regulated by means of a reservoir or dam, thereby facilitating a conversion of the water's potential energy to kinetic energy. To a greater extent than wind energy, hydroenergy is geographically dependent, being limited to locations where rivers are found. Hydroenergy is a natural consequence of the earth's water cycle, which

Figure 1.3
Wind farms generate electrical power on a commercial scale.

(Majeczka/Shutterstock)

is driven by the sun. Thus, hydroenergy, like wind energy, is an indirect form of solar energy.

Of all the renewable energy sources, hydro is the oldest. Power from water was used anciently for grinding grain, pumping water, and papermaking. Today, the primary use of water power is the generation of electricity. Of all the renewable energy sources, hydropower accounts for approximately three-fourths of the world's electricity. The power-generation capacities of hydropower systems vary widely, ranging from about 5 kW for running a few lights and appliances in a home to over 20 GW for powering large cities. Hydropower is covered in more detail in Chapter 4. A large hydropower plant is shown in Figure 1.4.

1.5 GEOTHERMAL ENERGY

Geothermal energy originates within the earth. About 80 percent of this internal energy is heat generated by the decay of radioactive substances, and the remainder is residual heat from the earth's planetary accretion. The energy from these two sources is transported to the earth's surface where it is manifested in hot ground and surface waters, rocks, and magma. Even though the heat content of the earth is finite, geothermal energy is considered to be renewable because the rate of heat extraction is very small compared with the earth's heat content, which is estimated to be 10^{31} J.

Like hydroenergy, geothermal energy has been utilized by humans for centuries. Heat from the earth has been used for heating living spaces and water since ancient times. In modern times, geothermal energy is still used for these applications but is also used for electrical power generation. In 2010 geothermal energy

Figure 1.4
The Three Gorges Dam, spanning the Yangtze River in China, is the world's largest hydropower plant with a generating capacity of 22.5 GW.

(Thomas Barrat/ Shutterstock)

accounted for approximately 10 GW of global electrical power generation, and it is projected to supply 1 percent of global electrical power generation by 2050. Unlike wind and hydroenergy, geothermal energy is not an indirect form of solar energy because it is not induced by the sun nor does it depend on the sun for its sustenance. A geothermal power plant is shown in Figure 1.5. Geothermal energy is discussed in more detail in Chapter 5.

1.6 MARINE ENERGY

Marine energy, often referred to as *ocean* energy, is a broad category that includes energy derived from oceanic tides as well as oceans at locations away from shorelines. *Tidal* energy is derived from the periodic rise and fall of sea levels, which is

Figure 1.5
The Don Campbell geothermal power plant, located in Mineral County, Nevada, has a power-generation capacity of 16 MW.

(Courtesy of Ormat Technologies, Inc.)

Figure 1.6
The La Rance Tidal Barrage, located on the estuary of the Range River in Brittany, France, was the world's first tidal power station. Its power-generation capacity is 240 MW.

(TSWGB)

caused by a combination of gravitational forces exerted by the moon and sun and the rotation of the earth. The potential or kinetic energy of the tides is converted to mechanical energy in a turbine and then electrical energy in a generator. *Ocean* energy is derived from three sources—ocean surface waves, underwater ocean currents, and temperature differences in ocean water. For the first two sources, kinetic energy of the waves and currents is converted to mechanical energy, which is then converted to electrical energy. For the third source, a heat engine utilizes a temperature difference of deep and shallow water to generate mechanical energy in a turbine and then electrical energy in a generator. As long as the earth's oceans remain in the liquid phase, which is equivalent to saying that as long as the sun maintains the earth's surface at suitable temperatures, marine energy will be available. Thus, marine energy, like wind and hydroenergy, is an indirect form of solar energy. Marine energy is covered in Chapter 6. A tidal power plant is shown in Figure 1.6.

1.7 ENERGY FROM BIOMASS

Biomass is the earth's living, or recently living, matter found within the biosphere, the thin atmospheric layer near the earth's surface. Energy stored in biomass is naturally recycled through a series of chemical and physical conversion processes in the soil and surrounding atmosphere, which is the reason that energy from biomass is considered renewable. A part of this energy can be captured by intervening at the right time when biomass is available as fuel. If the rate of consumption does not exceed the rate at which biomass is recycled, the combustion of biomass does not generate more heat or carbon dioxide than does natural processes. Biomass can be directly converted to energy by combusting it to heat a living space or to heat water to run a turbine for generating electricity. Alternatively, biomass can first be converted to a secondary substance that is subsequently combusted or combined with other substances prior to combustion. Biomass is a renewable energy source in the respect that plants can be quickly regrown to replace the plants that were consumed. Because biological organisms depend on the sun, solar energy is the ultimate source of energy from biomass.

Figure 1.7
Wood-fired power plant in northern California operated by Wheelabrator Shasta Energy Company.

(Wheelabrator Shasta Energy/NREL)

Humans have utilized biomass for heating and cooking for centuries, but the use of biomass for electrical power generation is a relatively new development. A power plant that utilizes biomass for fuel is shown in Figure 1.7. Biomass is covered in more detail in Chapter 7.

1.8 ENERGY AND POWER CALCULATIONS

To carry out an engineering analysis of a renewable energy system, we must understand how to do energy and power calculations. *Energy* is defined as the capacity to do work. If a device or system has the capacity to do work, it possesses at least one form of energy that is available for transformation to another form of energy. As a simple example, consider a boulder poised on the edge of a cliff, as illustrated in Figure 1.8. By virtue of the boulder's height, z, above the ground, the boulder has **potential energy**, *PE*, that is calculated using the formula

$$PE = mgz \tag{1.1}$$

where m is the boulder's mass and g is the acceleration of gravity ($g = 9.81$ m/s^2 = 32.2 ft/s^2). As the boulder falls to the ground, its potential energy is converted to **kinetic energy**, *KE*, that is calculated using the formula

$$KE = \tfrac{1}{2}\, mv^2 \tag{1.2}$$

where v is the boulder's velocity at a given point as the boulder falls. Immediately before the boulder impacts the ground, all of its potential energy has been converted to kinetic energy, in accordance with the **law of conservation of energy**, which states that energy is conserved in all energy transformations. This law is also known as the *first law of thermodynamics*.

Potential energy and kinetic energy are key quantities in the analysis of renewable energy systems, particularly wind, hydro, and marine systems. It is common,

Figure 1.8
As a boulder falls toward the ground, its potential energy is converted to kinetic energy.

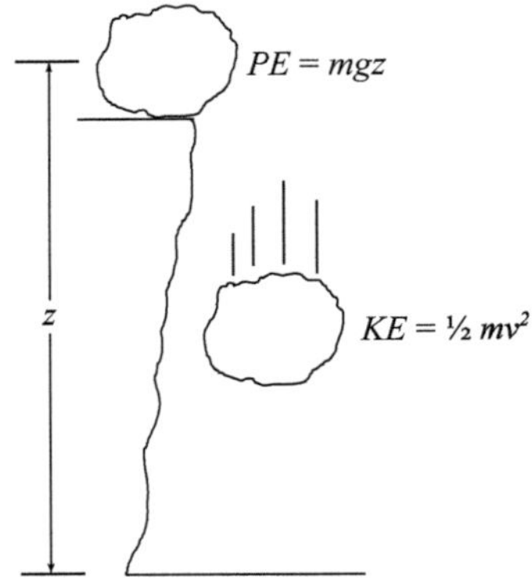

and mathematically convenient, to express potential energy and kinetic energy on a *per unit mass* basis by dividing these quantities by mass *m*, which yields

$$pe = gz \qquad (1.3)$$

$$ke = \tfrac{1}{2}\, v^2 \qquad (1.4)$$

Lowercase letters are typically used to distinguish energy per unit mass from energy.

Potential energy and kinetic energy are forms of *mechanical energy* that relate to the location and velocity, respectively, of a body or substance with respect to an outside reference frame. *Thermal, chemical,* and *nuclear* energy are forms of *internal energy* that relate to the microscopic behavior of a substance. **Thermal energy** is the energy associated with the internal kinetic energy of the molecules of a substance. This form of energy is manifested by the temperature of the substance. **Chemical energy** is the energy associated with atomic bonds in a molecule. Atomic bonds are broken or formed during chemical reactions such as combustion. **Nuclear energy** is the energy associated with the strong bonding forces between the particles in the nucleus of an atom. Nuclear energy is released when these particles break apart (fission) or combine (fusion). Thermal energy and chemical energy are key quantities in the analysis of solar, geothermal, marine, and biomass energy systems.

The unit for energy in the SI unit system is the joule (J). A joule is defined as a newton-meter,

$$1\,J = 1\,N{\cdot}m$$

where newton (N) is the unit of force and meter (m) is the unit of length in the SI unit system. One joule is a very small amount of energy, so units of kJ (10^3 J) or MJ (10^6 J) are typically used in calculations. One Btu is about the energy released by the complete burning of a wooden kitchen match. The conversion factor between the joule and Btu is

$$1\,Btu = 1055.06\,J$$

Another useful conversion factor for energy is

$$1\,kWh = 3600\,kJ$$

A kilowatt-hour (kWh) is the unit of energy typically used by electric power companies.

Power is the time rate of energy. Stated another way, power is energy divided by a time interval. The watt (W), the unit of power in the SI unit system, is defined as a joule per second,

$$1\,W = 1\,J/s$$

In the English unit system, horsepower (hp) is a commonly used unit for power. Other power units commonly used in the English unit system are Btu/h and Btu/s. Here are some useful conversion factors for power:

$$1\,W = 3.4121\,Btu/h = 3.6\,kJ/h \quad 1\,hp = 745.7\,W = 2544.4\,Btu/h$$

EXAMPLE 1.1

The boulder shown in Figure 1.8 has a mass of 100 kg, and the cliff is 50 m high. When the boulder is poised on the cliff, what is the potential energy of the boulder? If the boulder is pushed off the cliff, what is the boulder's velocity immediately before the boulder impacts the ground?

SOLUTION

Using Equation (1.1), the potential energy of the boulder is

$$PE = mgz$$
$$= (100\,kg)(9.81\,m/s^2)(50\,m)$$
$$= 49.1 \times 10^3\,kg{\cdot}m^2/s^2 = 49.1 \times 10^3\,J = 49.1\,kJ$$

You probably wonder how units of $kg{\cdot}m^2/s^2$ is equivalent to the unit of joule. The unit of joule must be broken down to its base units to see how this works. Remember, a joule is defined as a newton-meter. But, according to Newton's second law of motion, force is mass multiplied by acceleration. So, the newton is defined as a kilogram multiplied by meter per second squared,

$$1\,N = 1\,kg{\cdot}m/s^2$$

Hence, one joule is

$$1\,J = 1\,N{\cdot}m = 1\,(kg{\cdot}m/s^2){\cdot}m = 1\,kg{\cdot}m^2/s^2$$

To find the boulder's velocity immediately before the boulder impacts the ground, we must recognize that, according to the law of conservation of energy, all the potential energy of the boulder is converted to kinetic energy immediately before impact. Energy associated with aerodynamic drag forces is very small and can be neglected. Thus, we can equate potential energy to kinetic energy,

$$mgz = \tfrac{1}{2}\,mv^2$$

Dividing by mass, m, and solving for velocity, v, we obtain

$$v = \sqrt{2gz}$$
$$= \sqrt{2(9.81\,m/s^2)(50\,m)}$$
$$= 31.3\,m/s$$

EXAMPLE 1.2

The amount of solar radiation incident on a solar panel is 1620 W, but the panel absorbs only 15 percent of this radiation because of surface reflectance and panel efficiency factors. How much energy is absorbed by the panel during a two-hour period? Express the answer in units of MJ and Btu.

SOLUTION

The power absorbed by the solar panel is

$$P_{absorbed} = 0.15(1620\,\text{W})$$
$$= 243\,\text{W} = 243\,\text{J/s}$$

Power is energy divided by time, so we multiply absorbed power by the time interval, which is 2 h (7200 s). We obtain

$$E_{absorbed} = P_{absorbed}\,\Delta t$$
$$= (243\,\text{J/s})(7200\,\text{s})$$
$$= 1.75 \times 10^6\,\text{J} = 1.75\,\text{MJ}$$

Using the conversion factor between J and Btu, we obtain

$$1.75 \times 10^6\,\text{J} \times \frac{1\,\text{Btu}}{1055.06\,\text{J}} = 1.66 \times 10^3\,\text{Btu}$$

EXAMPLE 1.3

At a site being considered for a wind turbine, the wind blows steadily at 12 mi/h. Find the kinetic energy of the wind.

SOLUTION

We first convert the wind velocity of 12 mi/h to units of m/s. Using a conversion factor for velocity, we obtain

$$12\,\text{mi/h} \times \frac{1\,\text{m/s}}{2.2369\,\text{mi/h}} = 5.36\,\text{m/s}$$

Using Equation (1.4), the kinetic energy per unit mass of air is

$$ke = \tfrac{1}{2}\,v^2$$
$$= \tfrac{1}{2}\,(5.36\,\text{m/s})^2$$
$$= 28.7\,\text{m}^2/\text{s}^2 = 28.7\,\text{J/kg}$$

Using a similar breakdown of units as was used in Example 1.1, it can be shown that m^2/s^2 is equivalent to J/kg. To find the power-generation potential of a wind turbine at this site, the air density and turbine blade diameter would have to be known. Energy analysis of wind turbines is covered in Chapter 3.

EXAMPLE 1.4

Electrical power is to be generated by installing a hydropower plant at a reservoir. The turbine generator that converts potential energy of water into electrical energy is located 120 m below the surface of the water. Find the potential energy of the water for this hydropower plant.

SOLUTION

Using Equation (1.3), the potential energy per unit mass of water is

$$pe = gz$$
$$= (9.81\,\text{m/s}^2)(120\,\text{m})$$
$$= 1.18 \times 10^3\,\text{m}^2/\text{s}^2 = 1.18 \times 10^3\,\text{J/kg}$$

To find the power-generation potential of a hydropower plant that uses this reservoir, the mass flow rate of water through the turbine would have to be known. Energy analysis of hydropower plants is covered in Chapter 4.

Before leaving this section, let us define two additional technical terms—*efficiency* and *capacity factor*. When energy is transformed from one form to another, the useable output energy is always less than the input energy. This is because of energy losses that occur within the system. **Efficiency** is the *ratio of the useable output energy to the input energy*,

$$\text{efficiency} = \frac{\text{useable output energy}}{\text{input energy}} \tag{1.5}$$

Efficiency, which is a unitless quantity, is always less than 1 and is often expressed as a percentage. For example, the efficiency of an electrical generator is typically greater than 90 percent, whereas the efficiency of a coal-fired power plant is usually around 40 percent. Efficiency can also be expressed as the ratio of useable output power to input power.

In the design and cost analysis of a power plant, engineers use another useful quantity, the power plant **capacity factor**, *CF*, defined as the actual output energy of a plant for a given time period divided by the maximum possible output energy of the plant for the same time period,

$$CF = \frac{\text{actual output energy}}{\text{maximum possible output energy}} \tag{1.6}$$

Like efficiency, capacity factor is a unitless quantity and can also be expressed as a percentage. Typically, the time period for which capacity factor applies is one year, but other time periods can be used. There are 8760 hours in one year (365 day/year $\times$ 24 h/day), so, for example, a 1 MW power plant running constantly has a *CF* of 1 and generates 8760 MWh of energy per year. A 1 MW power plant realistically runs a portion of the year, so it might generate only 3750 MWh of energy, giving a capacity factor of (3750 MWh/8760 MWh) = 0.428 (42.8 percent).

Efficiency and capacity factor will be revisited in later chapters as each renewable type is discussed.

1.9 ECONOMICS

Renewable energy is more than a technology enterprise—it's a business, and all businesses are measured by economics. For renewable energy, there are basically three types of economics to consider: pecuniary, social, and physical. The term *pecuniary* refers to money and its relationship to business. The initial investment of building a hydroelectric plant is a pecuniary aspect of economics. *Social* economics refers to the costs that society bears. The health care cost of air pollution by burning fossil fuels is a social aspect of economics. *Physical* economics refers to the cost of energy and the efficiencies of the processes involved. Nature imposes fundamental limitations on all energy transformations, and we pay for those limitations. For example, there is only so much power available in the wind for a given wind speed and turbine blade diameter.

There are numerous factors to consider in the economics of renewable energy systems in commercial and residential applications. Some of the primary factors are installation cost, energy demand, cost of energy, size of energy system, operation and maintenance, variability of energy, reliability, inflation, depreciation, and incentives. A full economic analysis of a renewable energy system must take these factors, and others, into account. For our purposes, we will consider two simple methods for performing a basic economic analysis of a renewable energy system—simple payback and cost of energy.

Simple payback, also called *payback period*, refers to the period of time required to recoup the cost of an investment. For example, if a residential photovoltaic solar system costs $7000 to purchase and install, and the system saves the homeowner $1000 per year, the payback period is seven years because it took seven years for the system to reach a break-even point, that is, to pay for itself. If *operation and maintenance (O&M)* costs and the time value of money are neglected, *simple payback (SP)* in years is calculated using the formula,

$$SP = \frac{IC}{(AEP)(EC)} \tag{1.7}$$

where *IC* is *initial cost* in units of $, *AEP* is *annual energy production* in units of kWh/y, and *EC* is *energy cost* in units of $/kWh.

EXAMPLE 1.5

The initial cost of a wind turbine for a small business is $38,000. If the annual energy production of the turbine is 33,400 kWh/y and the energy cost (price of electricity charged by the electric utility) is $0.09/kWh, what is the simple payback?

SOLUTION

Using Equation (1.7), we have

$$SP = \frac{IC}{(AEP)(EC)}$$

$$SP = \frac{\$38,000}{(33,400 \text{ kWh/y})(\$0.09/\text{kWh})}$$

$$= 12.6 \text{ y}$$

A more accurate formula for simple payback that includes $O\&M$ and the time value of money is

$$SP = \frac{IC}{(AEP)(EC) - (IC)(FCR) - AOM} \tag{1.8}$$

where FCR is fixed charge rate in units of y^{-1}, and AOM is annual $O\&M$ cost in units of \$/y. The quantity FCR is typically the interest rate on a loan or the value of interest received if savings had not been used to pay part or all of the initial cost.

EXAMPLE 1.6

Rework Example 1.5 if a loan at 3.5 percent was used to finance half of the wind turbine and the annual operation and maintenance cost is \$100/y.

SOLUTION

The value of FCR is 0.035, and the value of IC in the denominator of Equation (1.8) is \$19,000, half the initial cost of the system. Therefore, the simple payback is

$$
\begin{aligned}
SP &= \frac{IC}{(AEP)(EC) - (IC)(FCR) - AOM} \\
&= \frac{\$38,000}{(33,400\ \mathrm{kWh/y})(\$0.09/\mathrm{kWh}) - (\$19,000)(0.035/\mathrm{y}) - \$100/\mathrm{y}} \\
&= 17.0\,\mathrm{y}
\end{aligned}
$$

As expected, the inclusion of $O\&M$ cost and the time value of money increased the payback period.

The **cost of energy** (COE) is the "levelized" or average value of the energy produced by a renewable energy system over its lifetime. The lifetime of a renewable energy system is a function of the type of system (solar, wind, geothermal, etc.). For photovoltaic systems, a typical lifetime is 30 years, whereas for wind turbines a typical lifetime is about 20 years. Some hydropower plants have been in operation for well over 50 years. The COE is a measure of the economic feasibility of the renewable energy system and is compared to the cost of energy from the local electric utility or other providers.

The major factors in the COE are installed cost and annual energy production, but levelized replacement cost (LRC) and annual fuel cost (AFC) influence COE as well. The units of both LRC and AFC are \$/y. The relation for COE is

$$COE = \frac{(IC)(FCR) + LRC + AOM + AFC}{AEP} \tag{1.9}$$

where, as in simple payback, AOM is annual $O\&M$ cost in units of \$/y.

EXAMPLE 1.7

A wind turbine with a power-generation capacity of 1.5 MW costs \$3.25 million to install. The fixed charge rate is 7 percent, the annual operation and maintenance cost is assumed to be 1 percent of the initial cost, and the levelized replacement cost is averaged over an expected 20-year lifetime. If the capacity factor for this turbine is 0.25, what is the cost of energy?

SOLUTION

The turbine has a power capacity of 1.5 MW, so the amount of energy that the turbine can produce in one year if it runs constantly is

$$E_{max} = (1.5\,\text{MW})\,(8760\,\text{h/y})$$
$$= 13{,}140\,\text{MWh/y}$$

The capacity factor is 0.25, so the actual annual energy production of the turbine is

$$AEP = (0.25)\,(13{,}140\,\text{MWh/y})$$
$$= 3285\,\text{MWh/y}$$

For a wind turbine, the annual fuel cost, AFC, is zero. The LRC and AOM are

$$LRC = (\$3.25 \times 10^6)/(20\,\text{y})$$
$$= \$162{,}500/\text{y}$$
$$AOM = (0.01/\text{y})\,(\$3.25 \times 10^6)$$
$$= \$32{,}500/\text{y}$$

Substituting values into Equation (1.9), we obtain

$$COE = \frac{(IC)\,(FCR) + LRC + AOM + AFC}{AEP}$$

$$COE = \frac{(\$3.25 \times 10^6)\,(0.07/\text{y}) + \$162{,}500/\text{y} + \$32{,}500/\text{y}}{3285\,\text{MWh/y}}$$

$$= \$128.6/\text{MWh}$$

which, when converted to units of \$/kWh, is approximately

$$COE = \$0.129/\text{kWh}$$

This cost of energy would have to be compared with the average cost of energy from the local utility over the next 20 years to determine whether the wind turbine system is economically viable.

1.10 ENVIRONMENTAL CONSIDERATIONS

Through farming, construction, and other activities, people have always changed their surroundings. In the decades following the industrial revolution, the consumption of coal steadily increased, resulting in severe pollution of major cities. In the twentieth century, coal consumption continued to increase, but because of the use of smokeless fuel and tall smokestacks, pollution in populated areas was significantly reduced. However, forests in regions of North America and Europe were dying, and marine life in lakes was dying as well. Scientists discovered that coal combustion was causing *acid rain*, a form of precipitation arising from emissions of sulfur dioxides and nitrogen oxides that react with water in the atmosphere.

Acid rain is not the only harmful effect of coal combustion. The combustion of coal and other fossil fuels releases carbon dioxide into the atmosphere. Carbon dioxide, along with water vapor, methane, nitrogen oxides, and ozone, is a greenhouse gas. In recent decades the surface temperature of the earth has markedly increased. Scientists have proposed that the accelerated use of fossil fuels is at least partly, and perhaps primarily, to blame for this temperature rise. If the combustion of fossil fuels is responsible for climate change and if we persist in using fossil fuels at current or accelerated rates, the earth's temperature might continue to rise, leading to gradual melting of polar ice caps and subsequent rises in sea level that will inundate coastal areas. Another likely effect of climate change is the increased frequency of extreme weather phenomena such as droughts and heavy rain fall. Climate change could even cause species extinctions due to shifting temperature patterns.

For practical and economic reasons, the consumption of fossil fuels cannot be suspended or even significantly slowed in the coming years. According to the U.S. EIA, approximately 83 percent of the total energy (electrical, transportation, heating, etc.) consumed in the United States in 2010 came from fossil fuels, about 9 percent came from nuclear sources, and approximately 8 percent came from renewable sources. In an attempt to reduce carbon dioxide emissions and become more energy independent, the United States and other industrialized nations are undergoing a very gradual shift from fossil fuels to renewable sources.

Compared to energy systems that utilize fossil fuels, renewable energy systems do not release a significant amount of harmful substances into the atmosphere. However, all renewable energy systems affect the environment in some way. For example, solar panels require a significant amount of space and compete for this space when installed on roof tops. Most photovoltaic solar panels incorporate silicon, a semiconductor material that is very energy intensive to produce. The production of silicon requires electricity, most of which is generated by fossil-fuel power plants. Wind turbines generate a low to moderate amount of aerodynamic and mechanical noise and may be visually distracting, and rotating turbine blades pose a threat to birds. Reservoirs for hydropower plants inundate land, disrupting the local biological ecosystem and possibly displacing people from their homes. These and other environmental impacts of renewable energy systems are discussed in the chapters devoted to each renewable energy type.

PROFESSIONAL SUCCESS—ARE YOU AN "ENVIRONMENTALIST"?

The terms *environmentalist* and *environmentalism* invoke a variety of opinions, perceptions, and even emotions. One definition of environmentalism is "a political and ethical movement that seeks to improve and protect the quality of the natural environment through changes to environmentally harmful human activities." Environmentalists are sometimes referred to using derisive names such as *greenie, tree hugger,* or *forest freak.* In the opinion of some, environmentalism tends to be defined along political lines, typically compartmentalizing environmentalists as liberals and antienvironmentalists as conservatives. Like any political movement, extreme versions of environmentalism exist. Some notable environmentalists include former vice president of the United States Al Gore, former president of the United States Theodore Roosevelt, Prince of Wales Prince Charles, photographer and writer Ansel Adams, and writer and philosopher Henry David Thoreau.

If environmentalism is a political and ethical movement, is it an engineering movement as well? If it is not, should it be? Consider the first fundamental canon of the National Society of Professional Engineers (NSPE) of the Professional Engineers Code of Ethics:

> Engineers, in the fulfillment of their professional duties, shall hold paramount the safety, health, and welfare of the public.

Does not this engineering canon subsume the objectives of environmentalism? Holding paramount the safety, health, and welfare of the public does not only apply to the performance of automobiles, structural integrity of bridges, and bandwidth of communication devices. The Professional Engineers Code of Ethics applies to all products and processes designed by engineers that impact the environment, and that encompasses everything. There is even an engineering discipline, *environmental engineering*, that specifically focuses on environmental aspects of engineering.

So, are you an environmentalist? The professional responsibilities of engineers cannot be disconnected or disassociated from environmentalism. In this respect, all engineers are environmentalists.

SUMMARY

Energy is the capacity to do work, and power is the time rate of energy. Energy and power are expressed in SI units of the joule (J) and the watt (W), respectively. Renewable energy is energy that comes from sources that are naturally replenished. Fossil fuels are not renewable energy sources because these fuels take a very long time to be replenished in the earth. There are three primary types of power plants: conventional thermal, nuclear, and renewable.

The types of renewable energy introduced in this chapter are solar, wind, hydro, geothermal, marine, and biomass. Solar energy emanates directly from the sun. Wind energy, an indirect form of solar energy, is derived by converting kinetic energy of moving air to electrical energy. Hydroenergy refers to energy derived from flowing or falling water. Geothermal energy emanates from within the earth. Marine energy includes energy derived from oceanic tides, currents, and temperature differences in ocean water. Biomass is earth's living, or recently living, matter found within the thin layer of atmosphere near the earth's surface.

Energy and power calculations involve the conversion of various types of energy to other forms of energy using the first law of thermodynamics, the law of conservation of energy. Efficiency and capacity factor are important quantities in the analysis of renewable energy systems.

Renewable energy is a business as well as a technology enterprise. A basic economic assessment of a renewable energy system can be performed by a simple payback or cost of energy analysis.

Fossil fuels cause a variety of environmental problems, most notably the release of harmful substances into the atmosphere. Most renewable energy technologies do not emit harmful substances into the atmosphere but impact the environment in other ways.

KEY TERMS

biomass	hydroenergy	renewable energy
capacity factor	kinetic energy	simple payback
chemical energy	law of conservation of	solar energy
cost of energy	energy	sustainability
efficiency	marine energy	thermal energy
energy	nuclear energy	wind
fossil fuels	potential energy	
geothermal energy	power	

SUGGESTED READING

ASHRAE Handbook–HVAC Applications, New York: American Society of Heating, Refrigerating and Air-conditioning Engineers, 2011.

Boyle, G., Ed., *Renewable Energy–Power for a Sustainable Future* 3rd Ed., Oxford: Oxford University Press, 2012.

Breeze, P., *Power Generation Technologies*, Burlington, MA: Elsevier, 2005.

Hagen, K.D., *Introduction to Engineering Analysis* 4th Ed., Upper Saddle River, NJ: Prentice Hall, 2014.

Zooba, A.F., and R.C. Bansal, Eds., *Handbook of Renewable Energy Technology*, Hackensack, NJ: World Scientific, 2011.

PROBLEMS

1.1 Calculate the kinetic energy of a 0.145 kg baseball with a velocity of 80 mi/h. Express the answer in units of kJ.

1.2 A satellite in orbit about the earth has a photovoltaic solar panel that faces directly into the sun for ten hours per day. How much solar energy is incident on the panel during this period of time if the panel's surface area is 5 m^2?

1.3 The sun radiates energy at a rate of 3.9×10^{26} W. How much energy has the sun radiated since the beginning of the Common Era?

1.4 A residential gas-fired furnace is rated at 150,000 Btu/h. Convert this heating rate to units of kW and MW.

1.5 If a 40-W incandescent light bulb transfers 90 percent of its power as heat instead of light, what is the rate of heat transfer from the light bulb in units of Btu/h?

1.6 The metabolic rate (body heat production rate) of an adult engaged in swimming is approximately 2140 Btu/h. Convert this heat production rate to units of kW.

1.7 A coal-fired power plant steadily generates 10 MW of electrical power. In units of kJ, how much electrical energy is generated during a 24-hour period?

1.8 The available power in wind is proportional to turbine blade diameter and the cube of wind speed. For a given wind turbine, if the available power is 3 kW for a wind speed of 4 mi/h, what is the available power for a wind speed of 12 mi/h?

1.9 The average solar heat flux just outside the earth's atmosphere is 1366 W/m^2. Convert this quantity to units of Btu/h·ft^2. In units of MJ, how much solar energy is intercepted by a 12 m^2 surface at this location during a time period of six hours?

1.10 For a time period of 30 minutes, the solar radiation incident on a photovoltaic solar panel is 825 W/m^2. Assuming that one-tenth of the solar radiation is converted to electrical energy, what is the required surface area of the panel if 1.6 kWh of electrical energy is required?

1.11 A biomass power plant uses wood chips for fuel. The heat of combustion (heat released by the complete burning of a given amount of fuel) of wood chips is approximately $9000 \text{ Btu/lb}_\text{m}$. Convert this quantity to units of kJ/kg. (Note: $1 \text{ kg} = 2.2046 \text{ lb}_\text{m}$.)

1.12 The hot water from a geothermal source well steadily transfers 60 kW of heat to a working fluid that passes through a turbine. In units of Btu, how much energy is transferred to the working fluid during a 48-hour period?

1.13 An underwater turbine powered by an ocean current generates 2 kW of electrical power. This turbine is one of 24 identical devices in an ocean power plant. Assuming round-the-clock operation, how much energy in units of kWh does this power plant generate in a 30-day period?

1.14 The available power in an ocean wave is proportional to the square of wave height. For a given ocean wave, if the available power is 3.8 MW for a wave height of 1.2 m, what is the available power for a wave height of 4.5 m?

1.15 The initial cost of a wind turbine for a residential application is $14,000. If the annual energy production of the turbine is 10,500 kWh/y, and the energy cost is $0.10/kWh, what is the simple payback?

1.16 The initial cost of a micro hydro system is $1200. If the annual energy production of the system is 1800 kWh/y, and the price of electricity charged by the electric utility is $0.095/kWh, what is the simple payback?

1.17 Rework Problem 1.15 if a loan at 4.0 percent was used to finance $5000 of the wind turbine and the annual operation and maintenance cost is $150/y.

1.18 A hydroelectric plant with a power-generation capacity of 12 MW costs $20 million to install. The fixed charge rate is 7.0 percent, the annual operation and maintenance cost is assumed to be 1.0 percent of the initial cost and the levelized replacement cost is averaged over an expected 40-year lifetime. If the capacity factor for this plant is 0.38, what is the cost of energy?

2 Solar Energy

Objectives

After reading this chapter, you will have learned:

- What solar energy is
- The basic types of solar energy systems
- How to calculate the total solar irradiance on a collector
- How to calculate solar insolation
- How to calculate the electrical output power of a photovoltaic solar panel
- How to optimize the tilt angle of a photovoltaic solar panel
- How to do a basic energy analysis of a solar thermal system
- The basic environmental issues with solar energy

2.1 INTRODUCTION

Solar energy is the most vital of all energy sources, for without it no life on earth could exist. With the exception of nuclear energy and geothermal energy, all energy sources, including fossil fuels, directly or indirectly depend on the sun. Fossil fuels are the products of photosynthesis in plant matter that, over millions of years, formed into coal, oil, and natural gas beneath the surface of the earth. Winds are driven by pressure differences in the atmosphere caused by uneven heating of the earth's surface by the sun. In the absence of solar energy, the earth's waters would be frozen, making energy from flowing water impossible. Similarly, all forms of marine energy would be impossible without solar energy to maintain water on the earth in the liquid phase. Energy from biomass would also be impossible without solar energy because biomass, like fossil fuels, depends on photosynthesis. Nuclear energy originates from fission processes of radioactive substances mined from the earth and does not depend on the sun for its sustenance. Geothermal energy originates within the earth primarily from the decay of radioactive substances and to a lesser extent from the residual heat from the earth's planetary accretion. The sun neither causes nor sustains these processes.

In this chapter we examine the magnitude of energy that is generated by the sun and its capacity for supplying the world's energy demands. For the purpose of solar collector analysis and design, we present a methodology for calculating the amount of solar energy that is intercepted by a surface of any orientation. Using this methodology, we demonstrate how to perform a basic energy analysis of photovoltaic and solar thermal systems. We conclude the chapter by discussing the environmental issues associated with solar energy.

2.2 THE SOLAR ENERGY SOURCE

Solar energy emanates from the sun, the yellow star at the center of our solar system. Like all stars, the sun generates vast amounts of energy by nuclear fusion reactions in its interior. The energy from these reactions is transported to the sun's surface where it radiates into space. The sun has been radiating energy at a reasonably steady rate for several billion years and is predicted to continue to do so for billions of years to come, which is the basis for considering solar energy as a renewable source.

The sun radiates energy at a rate of 3.9×10^{26} W. At the outer edge of the earth's atmosphere, the average solar heat flux (solar power intercepted by a plane surface facing directly into the sun) during the year is approximately 1366 W/m^2. This quantity, known as the *solar constant, I_{sc}*, is based on recent solar heat flux measurements and is subject to revision as new measurements are made. A portion of the solar radiation incident on the earth is transmitted completely through the atmosphere and is therefore usable for earth-bound solar energy systems. However, a portion of the solar radiation incident on the earth is reflected from the atmosphere to space and is therefore unusable as solar energy on the earth. The remaining portion of the solar radiation incident on the earth is absorbed by the atmosphere and interacts with atmospheric carbon dioxide, water vapor, and particulate matter. Some of this energy is reradiated to space, and the rest is transported to the earth's surface where it can be utilized. Thus, the power-per-square-meter input for all earth-based solar energy systems is significantly less than the solar constant.

Let us estimate the amount of solar power that reaches the surface of the earth by assuming that 50 percent of the solar radiation incident on the earth is usable. The average radius, R, of the earth is 6371 km, so the projected area (area of a circle), A, of the earth is

$$A = \pi R^2$$
$$= \pi (6371 \times 10^3 \, \text{m})^2$$
$$= 1.275 \times 10^{14} \, \text{m}^2$$

Hence, the solar power, $\dot{Q}_{\text{solar}}$, at the earth's surface is

$$\dot{Q}_{\text{solar}} = 0.5 A I_{sc}$$
$$= 0.5 (1.275 \times 10^{14} \, \text{m}^2)(1366 \, \text{W/m}^2)$$
$$= 8.71 \times 10^{16} \, \text{W}$$

Can solar power alone meet the world's demands? The global power consumption in 2010 has been estimated to be 1.7×10^{13} W. Dividing this number by the estimated solar power at the earth's surface, we obtain 1.95×10^{-4}. Multiplying this number by 100 to get a percentage, we obtain 0.0195 percent. Thus, only 0.0195 percent of the solar power at the earth's surface could theoretically satisfy the world's power demands.

Let us examine the capacity of solar energy another way. Based on the 2010 global power consumption, how much of the earth's surface, covered with solar panels, is required to satisfy the world's power demands? Using the projected area of the earth in our first calculation, the surface area required is

$$A_{\text{solar panel}} = (1.95 \times 10^{-4})(1.275 \times 10^{14}\,\text{m}^2)$$
$$= 2.49 \times 10^{10}\,\text{m}^2$$

which is approximately 9600 square miles, a land area about the size of Vermont.

Without taking into account energy conversion efficiency, energy storage, and distribution variables, these approximate but revealing calculations indicate that solar energy alone has the capacity to satisfy global energy demands.

2.2.1 Solar Energy Systems

Solar energy is associated with three energy conversion processes: *heliochemical, heliothermal,* and *helioelectrical.* Heliochemical energy conversion is the photosynthesis process, heliothermal energy conversion involves the conversion of solar radiation to heat, and helioelectrical energy conversion occurs in photovoltaic solar cells. The first conversion process, heliochemical, will be covered in more detail in Chapter 7 on biomass. In the case of the latter two processes, two types of solar energy systems are employed—**solar thermal** systems and **photovoltaic** systems. The first type of solar thermal system is typically designed to heat water for domestic hot water use, as shown in Figure 2.1, but could also be used to heat air for a living space or provide heat for other purposes. This system incorporates a *collector* that absorbs solar radiation, thereby heating a *working fluid* that is pumped through it. Working fluids in such systems are usually an ethylene-glycol mixture (antifreeze). The warm working fluid passes through a *heat exchanger* in a storage tank where the fluid's heat is transferred to the domestic water. The cool working fluid returns to the collector to be reheated by solar radiation. This type of system is typically found in residential and small commercial applications.

The second type of solar thermal system, commonly called a **concentrating solar power** plant, is designed to generate electrical power on a large scale. This type of system incorporates a *concentrator*, consisting of a mirror or an array of mirrors, that reflects sunlight and focuses it onto a very small area. As shown in Figure 2.2, parabolic dish and heliostat systems focus sunlight at a point, whereas

Figure 2.1
Solar water heating system.

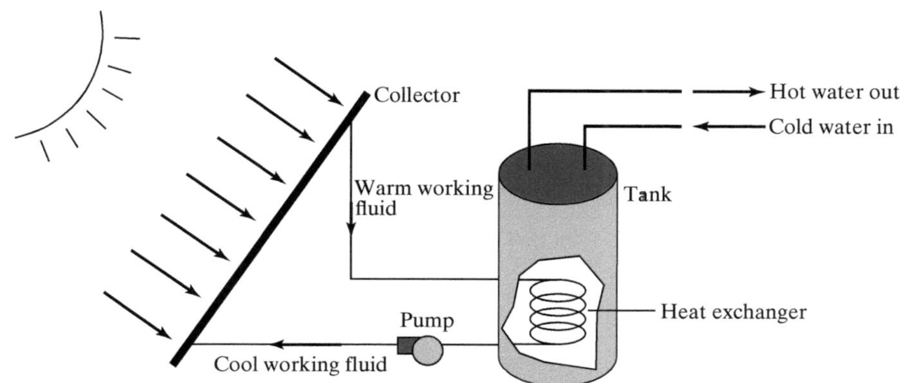

Figure 2.2
Concentrating solar power systems.

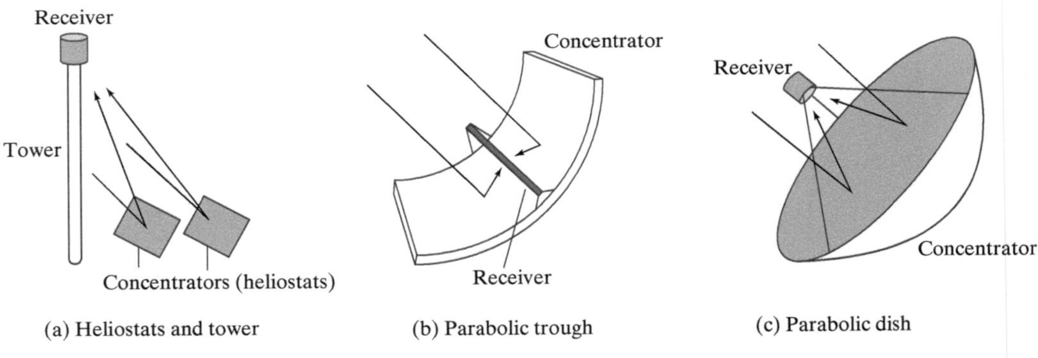

a parabolic trough system focuses sunlight along a line. For these concentrators to focus the sunlight properly, the concentrators must be able to track the sun as it moves across the sky. Dish and heliostat concentrators require dual-axis tracking, whereas a trough concentrator requires single-axis tracking. Located at the focus of the concentrator is a *receiver* that carries a working fluid, typically a molten salt mixture. Because the sunlit area of the receiver is very small, the heat flux (heating rate per unit surface area, measured in W/m^2) at the receiver is high enough to heat the working fluid to temperatures above 500°C. The hot working fluid passes through a heat exchanger where the fluid transfers heat to water, turning the water into steam. The rest of the system resembles a standard power plant in which steam drives a turbine, which in turn drives an electrical generator. Other components in the system include a condenser and pump, as shown in Figure 2.3.

Unlike a concentrating solar plant, a photovoltaic system converts light directly into electricity. The heart of a photovoltaic system is a *solar cell* constructed of silicon or other semiconductor material. The solar cell is arranged to form a *p-n* (positive-negative) junction such that when light strikes the cell, free electrons are created by the photoelectric effect. Under the influence of the junction's electric field, the electrons move through the cell to the cell's surface where they are collected by a metallic grid. The grid is connected to electrical contacts on both sides of the cell. If an external electrical circuit is connected across the contacts, a direct current (DC) electrical current flows through the circuit.

Figure 2.3
Concentrating solar power plant.

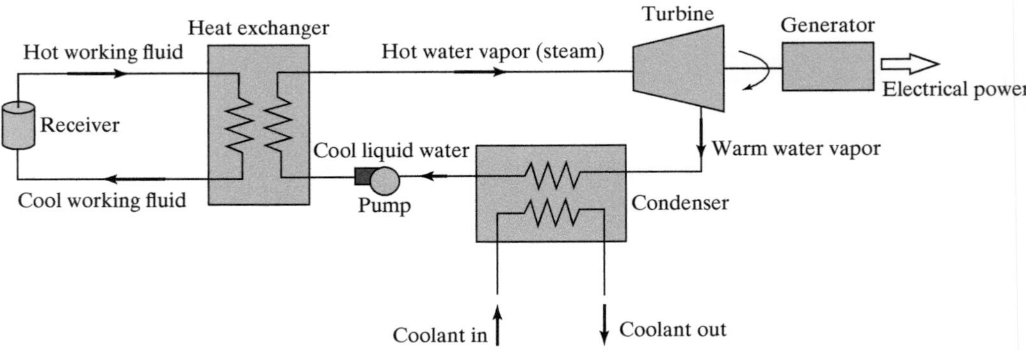

Figure 2.4
Photovoltaic hierarchy.

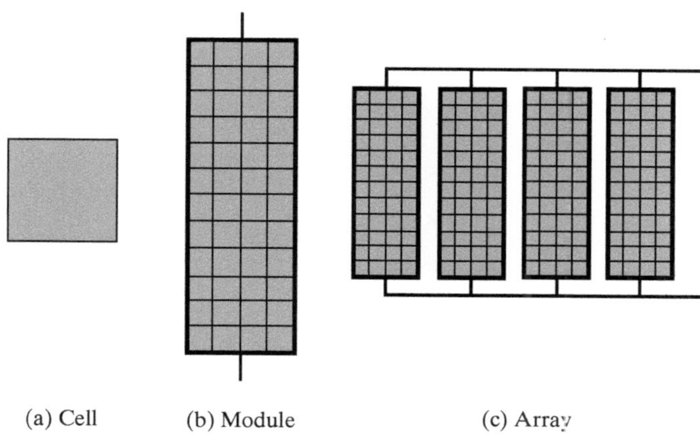

(a) Cell   (b) Module   (c) Array

A single solar cell produces an output voltage of approximately 0.5 V, which is insufficient to power most electrical devices. Hence, solar cells are connected in series, forming a *module*, to increase the output voltage. For example, a typical module designed to supply a standard 12-V output may consist of 30 or more cells to ensure reliable operation. To increase the output current, modules are connected in parallel, forming an *array* or *panel*. The desired voltage and current outputs can therefore be achieved by designing the array with the required number of series connections of cells and parallel connections of modules. This photovoltaic hierarchy is illustrated in Figure 2.4.

For residential and commercial building applications, photovoltaic systems are typically employed to augment electrical power provided by the local utility. In this *grid-connected* or *grid-tied* system, a solar panel and the power grid function in tandem to supply electrical power to the building, as shown in Figure 2.5. A charge controller prevents storage batteries from being overcharged and eliminates reverse current from the battery bank to the solar panel at night. The battery bank stores electrical energy generated by the solar panel during the day and uses that energy during the day or night, depending on the electrical power demands of the building. The inverter changes DC to alternating current (AC) to operate lights, appliances, and other electrical devices in the building. If the energy stored in the battery bank is insufficient to meet the entire electrical needs of the building, the power grid supplies the difference. Conversely, if the photovoltaic system generates more electrical energy than the building requires, the excess energy can be delivered to the power grid through the bidirectional power meter, thereby offsetting the electrical power the customer must purchase from the utility company at other times. This process is known as *net metering* and is a straightforward and low-cost way for the consumer to invest in renewable energy.

2.2.2 Solar Energy Calculations

To design or size the collector or concentrator of a solar thermal system or the solar panel of a photovoltaic system, it is necessary to calculate the solar energy incident on a surface. In this section we describe the computational procedure for

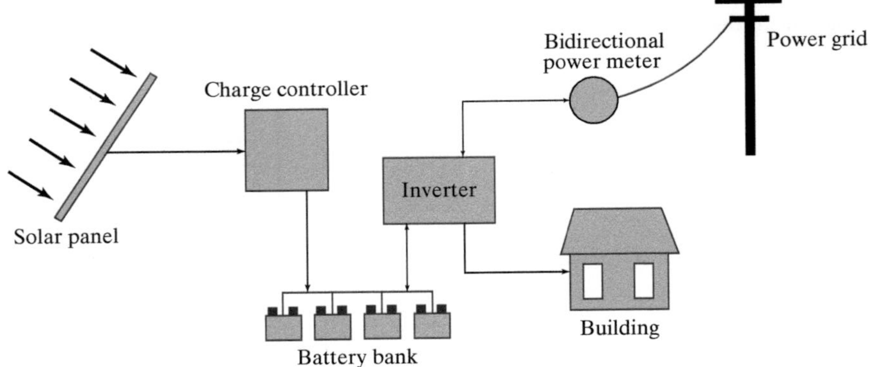

calculating the solar energy incident on a collection surface as a function of latitude, longitude, day of the year, time of day, surface orientation, and atmospheric conditions. The objective of such calculations is to orient the surface to maximize the amount of solar radiation collected and thus the amount of thermal or electrical energy that can be generated by the system.

2.2.2.1 Time and Solar Angles The earth completes one orbit about the sun every 365 days, and its orbital path is slightly elliptical. The imaginary plane of the earth's elliptical orbit is called the *ecliptic plane*. At perihelion on January 3, the earth is closest to the sun at a distance of approximately 147.1×10^6 km, and at aphelion on July 4, the earth is farthest from the sun at a distance of approximately 152.1×10^6 km. The earth spins on its own axis, rotating counterclockwise when viewed from above the north pole, completing one revolution in 24 hours. The earth's axis is tilted at 23.45° with respect to the ecliptic plane, so the northern hemisphere tilts toward the sun during summer and away from the sun during winter. The earth's orbit around the sun and the earth's tilt with respect to the ecliptic plane cause variations in the intensity and duration of solar radiation at different places on the earth, giving rise to the seasons and the apparent motion of the sun across the sky.

Declination is the angle of the sun north or south of the earth's equator. The declination angle, δ, for the northern hemisphere is approximated by the relation

$$\delta = 23.45° \sin\left[\frac{N + 284}{365} \times 360°\right] \tag{2.1}$$

where N is the day number of year ($N = 1$ for January 1 and $N = 365$ for December 31). For the southern hemisphere, the sign on δ is reversed. The surface of the earth is divided into a grid consisting of lines of longitude and latitude. The origin of 0° for longitude is the prime meridian, a north-south line that passes through Greenwich, England, and the origin of 0° for latitude is the equator.

Recognizing that there are 24 hours per day and 60 minutes per hour, there are 1440 minutes per day. This means that it takes four minutes (1440 min/360°) to traverse one degree of longitude. The *apparent solar time, AST*, also known as local solar time, for western longitudes is calculated from the equation

$$AST = LST + (4 \text{ min/deg})(LSTM - Long) + ET \tag{2.2}$$

where *LST* is local standard time for a particular time zone, *Long* is longitude, and *LSTM* is local standard time meridian. To adjust for daylight saving time, *LST* = *DST* − 1 h. *AST* is expressed in units of minutes. Values of *LSTM* for the seven standard time meridians in North America are given in Table 2.1. The quantity *ET* is called the *equation of time*, a measure of the extent by which solar time, as determined by a sundial, runs faster or slower than a clock running at a uniform rate. The equation of time is approximated by the relation

$$ET = 9.87 \sin(2D) - 7.53 \cos(D) - 1.5 \sin(D) \tag{2.3}$$

where

$$D = \frac{(N - 81)}{365} \, 360^{\circ} \tag{2.4}$$

Before discussing solar angles, let us work an example that illustrates the use of the foregoing equations.

Table 2.1 Values of *LSTM* for North American Time Zones

Time zone	LSTM
Atlantic	60°
Eastern	75°
Central	90°
Mountain	105°
Pacific	120°
Yukon	135°
Alaska-Hawaii	150°

EXAMPLE 2.1

Consider Ogden, Utah, on October 31 at 1:00 pm. Find the declination angle and apparent solar time.

SOLUTION

Ogden is in the mountain time zone, so *LSTM* = 105°. The longitude of Ogden is 111.97° west, and October 31 is day 304. From Equation (2.1), the declination angle is

$$\delta = 23.45^{\circ} \sin\left[\frac{(N + 284)}{365} \times 360^{c}\right]$$

$$= 23.45^{\circ} \sin\left[\frac{304 + 284}{365} \times 360^{c}\right]$$

$$= -15.06^{\circ}$$

(*continued*)

The local standard time is 1:00 pm, which means that 13 hours have elapsed since midnight. Hence, the local standard time, LST, is $13 \text{ h} \times 60 \text{ min/h} = 780 \text{ min}$. From Equation (2.4),

$$D = \frac{(N - 81)}{365} 360°$$

$$= \frac{(304 - 81)}{365} 360°$$

$$= 219.9°$$

From Equation (2.3), the equation of time, ET, is

$$ET = 9.87 \sin(2D) - 7.53 \cos(D) - 1.5 \sin(D)$$
$$= 9.87 \sin[(2)(219.9°)] - 7.53 \cos(219.9°) - 1.5 \sin(219.9°)$$
$$= 16.45 \text{ min}$$

From Equation (2.2), the apparent solar time, AST, is

$$AST = LST + (4 \text{ min/deg})(LSTM - Long) + ET$$
$$= 780 \text{ min} + (4 \text{ min/deg})(105° - 111.97°) + 16.45 \text{ min}$$
$$= 768.6 \text{ min}$$

Thus, the apparent solar time differs from the local standard time by 11.4 min. In the material that follows, the relationship between apparent solar time and solar angles is given.

The angles required for defining the position of the sun at any time are shown in Figure 2.6. The *hour angle, H,* is the azimuth angle of the sun's rays that changes as the earth rotates and is related to apparent solar time by the equation

$$H = \frac{AST - 720 \text{ min}}{4 \text{ min/deg}} \tag{2.5}$$

The hour angle is negative in the morning and positive in the afternoon, so $H = 0°$ at noon, apparent solar time. The solar altitude angle, β_1, is the apparent angular position of the sun if you are directly facing it. Solar altitude angle is calculated using the relation

$$\sin(\beta_1) = \cos(L) \cos(\delta) \cos(H) + \sin(L) \sin(\delta) \tag{2.6}$$

where L is latitude (taken to be positive in either hemisphere) and δ is declination angle. The solar azimuth angle, α_1, is the angular position of the sun from the north-south line and is related to the other angles by the equation

$$\cos(\alpha_1) = \frac{\sin(\beta_1) \sin(L) - \sin(\delta)}{\cos(\beta_1) \cos(L)} \tag{2.7}$$

The solar azimuth angle is negative in the morning and positive in the afternoon, so the sign of α_1 matches that of the hour angle, H.

Figure 2.6
Solar angles.

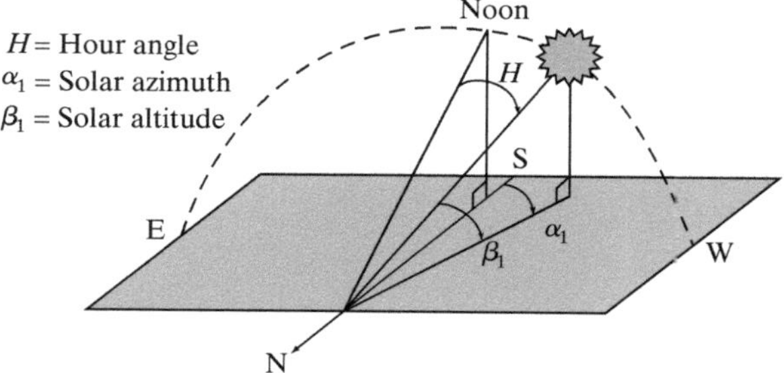

2.2.2.2 Collector Angles *Collector* is the generic term that we will use for the collector or concentrator of a solar thermal system or the solar panel of a photovoltaic system. Figure 2.7 shows a tilted collector surface and the pertinent angles. The quantity α_1 is the solar azimuth angle, and the quantity α_2 is the azimuth angle of the normal to the collector surface. For collectors facing east and west, $\alpha_2 = -90°$ and $+90°$, respectively. For collectors facing due south, $\alpha_2 = 0$. The quantity β_2 is the collector tilt angle (angle of collector surface with respect to the ground), and θ is the collector angle (angle between the sun and the normal to the collector surface). The collector angle is calculated using the equation

$$\cos(\theta) = \sin(\beta_1)\cos(\beta_2) + \cos(\beta_1)\sin(\beta_2)\cos(\alpha_1 - \alpha_2) \tag{2.8}$$

Note that if $\theta > 90°$, the sun is behind the collector, so the collector shades itself.

2.2.2.3 Solar Irradiance The total solar energy incident upon a surface is called **total solar irradiance**, I_{tot}, and is expressed in units of W/m^2. As illustrated in Figure 2.8, this quantity consists of three components—*direct* irradiance, I_D; *scattered* irradiance, also called sky radiation, I_S; and *reflected* irradiance, I_R. Total solar irradiance is the sum of these three components, so

$$I_{tot} = I_D + I_S + I_R \tag{2.9}$$

Figure 2.7
Solar and collector angles.

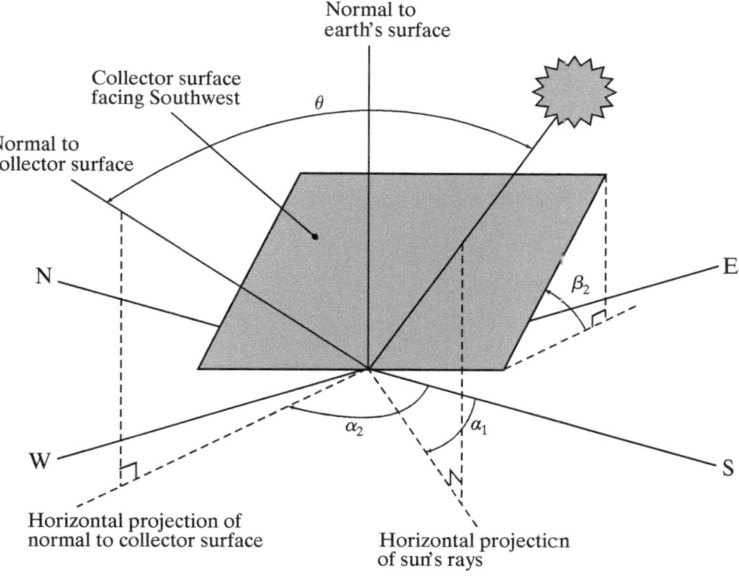

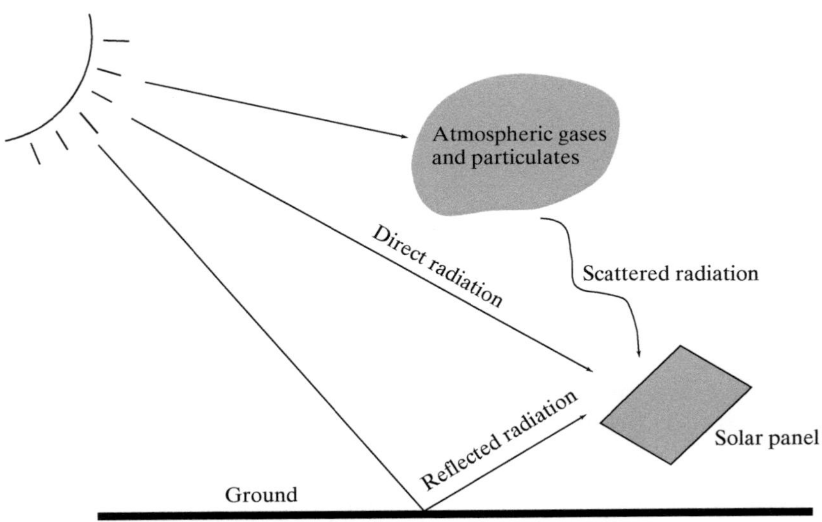

Direct irradiance is calculated by the equation

$$I_D = I_{DN} \cos(\theta) \tag{2.10}$$

where I_{DN}, the direct normal irradiance, is found using the relation

$$I_{DN} = A \exp\left(-\frac{p}{p_0}\frac{B}{\sin(\beta_1)}\right) \tag{2.11}$$

where β_1 is solar altitude and p/p_0 is the atmospheric pressure relative to a standard atmosphere, given by the equation

$$p/p_0 = \exp(-0.1184\,z) \tag{2.12}$$

where z is elevation in kilometers above sea level. The quantity A is the apparent extraterrestrial solar intensity, and B is the atmospheric extinction coefficient. Values of these two quantities are given in Table 2.2 for the 21st day of each month.

Table 2.2 Apparent Solar Irradiation, A, Atmospheric Extinction Coefficient, B, and Ratio of Diffuse Radiation on a Horizontal Surface to Direct Normal Irradiation, C, in the Northern Hemisphere

Date	Day of year (N)	A (W/m^2)	B	C
Jan 21	21	1230	0.142	0.058
Feb 21	52	1215	0.144	0.060
Mar 21	80	1186	0.156	0.071
Apr 21	111	1136	0.180	0.097
May 21	141	1104	0.196	0.121
Jun 21	172	1088	0.205	0.134
Jul 21	202	1085	0.207	0.136
Aug 21	233	1107	0.201	0.122
Sep 21	264	1152	0.177	0.092
Oct 21	294	1193	0.160	0.073
Nov 21	325	1221	0.149	0.063
Dec 21	355	1234	0.142	0.057

Table 2.3 Typical Foreground Reflectivity Values

ρ	Type of foreground
0.2	Ordinary ground or vegetation
0.8	Snow cover
0.15	Gravel
0.3	Concrete

Scattered irradiance is given by the equation

$$I_S = CI_{DN}\left[\frac{1 + \cos(\beta_2)}{2}\right] \tag{2.13}$$

where β_2 is collector tilt angle and C is the ratio of diffuse irradiation on a horizontal surface to the direct normal irradiation. Values of C are given in Table 2.2 for the 21st day of each month. For other days of the year, linear interpolation may be used to approximate A, B, and C.

Reflected irradiance is calculated using the equation

$$I_R = I_{DN}\,\rho\,(C + \sin(\beta_1))\left[\frac{1 + \cos(\beta_2)}{2}\right] \tag{2.14}$$

where ρ is foreground reflectivity and I_{DN} is given by Equation (2.11). Typical values of ρ for common surfaces are given in Table 2.3.

The total solar irradiance given by the foregoing equations applies only to clear atmospheric conditions. If the sky is cloudy or overcast, additional information would have to be known to reduce the irradiance accordingly.

2.2.2.4 Computational Procedure   The computational procedure for calculating the solar irradiance on a collector surface involves the use of Equations (2.1) through Equation (2.14). The following example serves to illustrate the calculations.

EXAMPLE 2.2

A collector with a tilt angle of 60° is located in Boston, Massachusetts, longitude 71.06°, latitude 42.36°, and elevation 43 m. The collector faces due south, and the foreground is grass. Find the total solar irradiance on the collector at 2:00 pm on February 21.

SOLUTION

February 21 is day 52 of the year, so the declination angle, using Equation (2.1), is

$$\delta = 23.45^{\circ} \sin\left[\frac{N + 284}{365} \times 360^{\circ}\right]$$

$$= 23.45^{\circ} \sin\left[\frac{52 + 284}{365} \times 360^{\circ}\right]$$

$$= -11.23^{\circ}$$

(continued)

From Equation (2.4),

$$D = \frac{(N - 81)}{365}\, 360°$$

$$= \frac{(52 - 81)}{365}\, 360°$$

$$= -28.61°$$

From Equation (2.3), the equation of time is

$$ET = 9.87 \sin(2D) - 7.53 \cos(D) - 1.5 \sin(D)$$
$$= 9.87 \sin[(2)(-28.61°)] - 7.53 \cos(-28.61°) - 1.5 \sin(-28.61°)$$
$$= -14.19 \text{ min}$$

Boston is in the eastern time zone, so $LSTM = 75°$. A local standard time of 2:00 pm gives

$$LST = 14\,\text{h} \times 60\,\text{min/h} = 840\,\text{min}$$

Using Equation (2.2), the apparent solar time is

$$AST = LST + (4\,\text{min/deg})(LSTM - Long) + ET$$
$$= 840\,\text{min} + (4\,\text{min/deg})(75° - 71.06°) + (-14.19\,\text{min})$$
$$= 841.6\,\text{min}$$

From Equation (2.5), the hour angle is

$$H = \frac{AST - 720\,\text{min}}{4\,\text{min/deg}}$$

$$= \frac{841.6\,\text{min} - 720\,\text{min}}{4\,\text{min/deg}}$$

$$= 30.40°$$

The solar altitude angle is calculated using Equation (2.6)

$$\sin(\beta_1) = \cos(L)\cos(\delta)\cos(H) + \sin(L)\sin(\delta)$$
$$= \cos(42.36°)\cos(-11.23°)\cos(30.40°) + \sin(42.36°)\sin(-11.23°)$$
$$= 0.4939$$

Therefore,

$$\beta_1 = \sin^{-1}(0.4939) = 29.60°$$

The solar azimuth angle is calculated using Equation (2.7),

$$\cos(\alpha_1) = \frac{\sin(\beta_1)\sin(L) - \sin(\delta)}{\cos(\beta_1)\cos(L)}$$

$$= \frac{\sin(29.60°)\sin(42.36°) - \sin(-11.23°)}{\cos(29.60°)\cos(42.36°)}$$

$$= 0.8211$$

Therefore,

$$\alpha_1 = \cos^{-1}(0.8211) = 34.80°$$

The collector faces due south, so $\alpha_2 = 0$. Using Equation (2.8), the cosine of the collector angle is

$$
\begin{aligned}
\cos(\theta) &= \sin(\beta_1) \cos(\beta_2) + \cos(\beta_1) \sin(\beta_2) \cos(\alpha_1 - \alpha_2) \\
&= \sin(29.60^\circ) \cos(60^\circ) + \cos(29.60^\circ) \sin(60^\circ) \cos(34.80^\circ - 0) \\
&= 0.8635
\end{aligned}
$$

The elevation of Boston is $z = 43$ m (0.043 km), so from Equation (2.12) we have

$$
\begin{aligned}
p/p_0 &= \exp(-0.1184\,z) \\
&= \exp[-(0.1184\,z)(0.043\ \text{km})] \\
&= 0.9949
\end{aligned}
$$

From Table 2.2 we see that for February 21, $A = 1215$ W/m^2, $B = 0.144$, and $C = 0.060$. Hence, using Equation (2.11), the direct normal irradiance is

$$
\begin{aligned}
I_{\text{DN}} &= A \exp\left(-\frac{p}{p_0}\frac{B}{\sin(\beta_1)}\right) \\
&= (1215\ \text{W/m}^2) \exp\left[\frac{(-0.9949)(0.144)}{\sin(29.60^\circ)}\right] \\
&= 909.1\ \text{W/m}^2
\end{aligned}
$$

Using Equation (2.10), the direct irradiance is

$$
\begin{aligned}
I_D &= I_{\text{DN}} \cos(\theta) \\
&= (909.1\ \text{W/m}^2)(0.8653) \\
&= 786.6\ \text{W/m}^2
\end{aligned}
$$

Scattered irradiance, given by Equation (2.13), is

$$
\begin{aligned}
I_S &= CI_{\text{DN}}\left[\frac{1 + \cos(\beta_2)}{2}\right] \\
&= (0.060)(909.1\ \text{W/m}^2)\left[\frac{1 + \cos(60^\circ)}{2}\right] \\
&= 40.91\ \text{W/m}^2
\end{aligned}
$$

From Table 2.3, the foreground reflectivity for grass is $\rho = 0.2$. Using Equation (2.14) the reflected irradiance is

$$
\begin{aligned}
I_R &= I_{\text{DN}}\rho(C + \sin(\beta_1))\left[\frac{1 + \cos(\beta_2)}{2}\right] \\
&= (909.1\ \text{W/m}^2)(0.2)[0.060 + \sin(29.60^\circ)]\left[\frac{1 - \cos(60^\circ)}{2}\right] \\
&= 25.18\ \text{W/m}^2
\end{aligned}
$$

Finally, the total irradiance is calculated using Equation (2.9),

$$
\begin{aligned}
I_{\text{tot}} &= I_D + I_s + I_R \\
&= (786.6 + 40.91 + 25.18)\ \text{W/m}^2 \\
&= 852.7\ \text{W/m}^2
\end{aligned}
$$

The computational procedure, illustrated by Example 2.2, is somewhat long and time consuming. To facilitate the speed of the calculations, the student is urged to incorporate the computational procedure into a spreadsheet or other computer-based tool.

An important quantity in the design and analysis of solar energy systems is **insolation**. Sometimes referred to as *solar irradiation* or *solar radiant exposure*, insolation is the incident solar energy per unit surface area. Insolation, H, is found by integrating the total solar irradiance over a specified time period. Thus,

$$H = \int_{t_1}^{t_2} I_{\text{tot}}\, dt \qquad (2.15)$$

where the time period is typically one day. This quantity, expressed in units of J/m^2, is the daily total solar energy that is incident on a unit surface area of a collector. Hence, the daily total solar energy that is incident on the collector is the insolation, H, multiplied by the surface area of the collector.

EXAMPLE 2.3

The collector in Example 2.2 has a surface area of $24\ m^2$. Find the total solar power incident on the collector at 2:00 pm. Also, find the insolation and the total solar energy incident on the collector during the day.

SOLUTION

The total solar power incident on the collector at 2:00 pm is the total solar irradiance multiplied by the surface area of the collector,

$$\dot{Q}_{\text{solar}} = I_{\text{tot}} A_{\text{collector}}$$
$$= (852.7\ \text{W/m}^2)(24\ \text{m}^2)$$
$$= 2.046 \times 10^4\ \text{W} = 20.46\ \text{kW}$$

To find the insolation, we must first calculate the total solar irradiance for the collector at several times throughout the day. Listed in Table 2.4 are the total solar

Table 2.4 Total Solar Irradiances for Example 2.3

Time	I_{tot} (W/m^2)
6:45 am	0
7:00 am	41.9
8:00 am	392
9:00 am	663
10:00 am	861
11:00 am	983
12:00 noon	1023
1:00 pm	979
2:00 pm	853
3:00 pm	651
4:00 pm	375
5:00 pm	26.1
5:15 pm	0

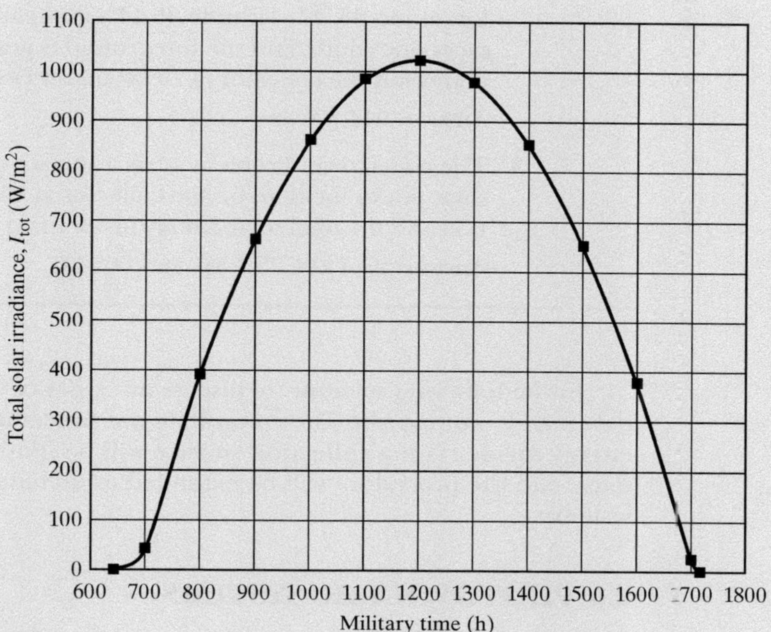

Figure 2.9
Total solar irradiance graph for Example 2.3.

irradiance values at one-hour time intervals from sunrise (about 6:45 am) to sundown (about 5:15 pm). A graph of these values, shown as a smooth curve drawn through the total solar irradiance data points, is shown in Figure 2.9.

Insolation, H, given by Equation (2.15), is the area under the curve in Figure 2.9. This area may be approximated using a numerical integration technique, the trapezoidal rule. This technique involves dividing the area under the curve into thin trapezoidal-shaped areas and adding them together to obtain the total area. The trapezoidal rule is explained in detail in Appendix A. Using the trapezoidal rule, the approximate insolation is

$$H = 24.6 \, \text{MJ/m}^2$$

The total solar energy incident on the collector during this day is the insolation multiplied by the surface area of the collector,

$$
\begin{aligned}
E_{\text{solar}} &= H \, A_{\text{collector}} \\
&= (24.6 \, \text{MJ/m}^2)(24 \, \text{m}^2) \\
&= 590.4 \, \text{MJ}
\end{aligned}
$$

PRACTICE!

1. Find the declination angle and apparent solar time for Richmond, Virginia, longitude 77.47°, latitude 37.53°, on September 21 at 2:00 pm.

   Answer: −0.202°, 837.8 min

2. A collector with a tilt angle of $50°$ is located in Nashville, Tennessee, longitude $86.78°$, latitude $36.17°$, and elevation 182 m. The collector faces due south, and the foreground is gravel. Find the total solar irradiance on the collector at 10:30 am on December 21.

   Answer: 914.3 W/m^2

3. The collector in Problem 2 has a surface area of 30 m^2. Find the total solar power incident on the collector at 10:30 am. Also, find the insolation and the total solar energy incident on the collector during the day.

   Answer: 27.4 kW, 22.1 MJ/m^2, 663 MJ

In the following sections we discuss two types of solar energy systems—*photovoltaic* and *solar thermal*. The computational procedure for calculating the solar energy incident on a collection surface will be illustrated for both types of systems, and the procedure will be extended to include the output power of these systems.

2.3 PHOTOVOLTAIC SYSTEMS

A photovoltaic system converts solar energy directly into electrical energy, which means that no intermediate energy transformations are involved. Solar radiation is converted directly to a DC voltage that can be used to power DC devices or can be inverted to AC voltage to power lights and appliances. The heart of a photovoltaic system is the solar cell, a device constructed of silicon or other semiconductor material. The solar cell is arranged to form a *p-n* (positive-negative) junction such that when light strikes the cell, free electrons are created by the photoelectric effect, a phenomena attributed to the French physicist Edmund Becquerel (1820–1891), who in 1839 discovered that electricity is produced when light strikes electrodes immersed in an electrically conductive solution. This phenomena was not universally accepted by the scientific community until Albert Einstein (1879–1955) published a paper on the photoelectric effect in 1904.

Development of the solar cell parallels that of the diode, transistor, and microchip. The solar cell was developed at Bell Laboratories in the 1950s and soon thereafter found applications in power generation for spacecraft. Solar cells remain an important technology for spacecraft electrical power generation, as shown in Figure 2.10. Since that time the solar cell has been improved and refined, largely due to advancements in materials science, finding applications in more common power-generation systems.

Because of a number of energy losses, a solar cell does not convert all of the incident solar power to electrical power. A portion of the surface of a solar cell is covered with a metallic grid to collect electrons generated by the photoelectric effect, rendering that portion of the cell ineffective as a collector. Some of the incident solar radiation on a solar cell is reflected from the front surface and therefore not utilized by the cell. Solar radiation consists of many wavelengths and therefore many energy levels, and some of the incident solar radiation does not have sufficient energy to initiate the photoelectric effect. On the other hand, some of the incident solar radiation has more than sufficient energy to initiate

the photoelectric effect, resulting in excess energy that is dissipated as heat in the semiconductor. Only a fraction of the photons with the appropriate amount of energy to eject an electron from its shell strikes an electron and ejects it. This fraction is known as quantum efficiency. There are two absorption-related energy losses. The first loss is due to electrons that are ejected from their shell and absorbed by impure atoms in the semiconductor. The second loss is due to photon absorption that occurs far from a p-n junction. These photons create electron hole pairs that immediately recombine, generating no electrical energy but a small amount of heat. The last type of energy loss is not related to the solar cell itself but its surroundings. If the solar cell is shaded by natural or human-made structures, the cell's output voltage decreases.

Taken together, the energy losses of a solar cell contribute to the **efficiency** of a solar cell. Solar cell efficiency, η_{cell}, is defined as the ratio of the electrical power generated by the cell to the solar power incident on the cell,

$$\eta_{\text{cell}} = \frac{P_{\text{elect}}}{\dot{Q}_{\text{solar}}} \tag{2.16}$$

The typical range of η_{cell} is approximately 0.15 to 0.25 (15 to 25 percent), but cell efficiencies of 0.40 and higher have been achieved. Solar cell efficiency can also be expressed in terms of a ratio of electrical energy produced by the cell to the solar energy incident on the cell during a specified time period,

$$\eta_{\text{cell}} = \frac{E_{\text{elect}}}{E_{\text{solar}}} \tag{2.17}$$

EXAMPLE 2.4

A commercial building in San Diego, California, has a roof-mounted photovoltaic solar panel that measures 8 m × 20 m. The panel is horizontal and has an efficiency of 0.21. How much electrical energy does the panel generate on November 21?

SOLUTION

The latitude and longitude of San Diego are 32.72° and 117.16°, respectively. From Table 2.2, $N = 325$, $A = 1221$ W/m^2, $B = 0.149$, and $C = 0.063$. San Diego is in the pacific time zone, so $LSTM = 120°$. A few of the quantities we need for solving this problem are not a function of time of day, so we calculate those first. From Equation (2.1), the declination angle is

$$\delta = -20.44°$$

and from Equation (2.4),

$$D = 240.7$$

so the equation of time, calculated from Equation (2.3), is

$$ET = 13.42 \text{ min}$$

The panel is horizontal ($\beta_2 = 0°$), so the reflected solar irradiance, I_R, given by Equation (2.14), is zero. San Diego is a sea-level city ($z = 0$), so according to Equation (2.12), $p/p_0 = 1$.

The rest of the quantities are a function of time of day and are summarized in Table 2.5. Time of day is expressed in military time. Note that the order of the quantities along the rows in this table corresponds with the computational order of the quantities themselves.

A graph of the total solar irradiance, shown in Figure 2.11, is constructed using the results in the last column of Table 2.5. Using the trapezoidal rule, we find the insolation to be

$$H = 13.78 \text{ MJ/m}^2$$

Table 2.5 Summary of calculations for Example 2.4

Time (h)	AST (min)	H (deg)	β_1 (deg)	α_1 (deg)	θ (deg)	I_{DN} (W/m^2)	I_D (W/m^2)	I_S (W/m^2)	I_R (W/m^2)	I_{tot} (W/m^2)
645	429.8	−72.55	2.73	−63.50	87.27	53.3	2.54	3.60	0	6.14
700	444.8	−68.80	5.52	−61.37	84.48	259.7	25.0	16.4	0	41.4
800	504.8	−53.80	16.07	−51.90	73.93	712.7	197.3	44.9	0	242.2
900	564.8	−38.80	25.19	−40.46	64.81	860.3	366.1	54.2	0	420.3
1000	624.8	−23.80	32.17	−26.54	57.83	923.0	491.5	58.1	0	549.6
1100	684.8	−8.80	36.18	−10.23	53.82	948.6	559.9	59.8	0	619.7
1200	744.8	6.20	36.51	7.23	53.49	950.5	565.5	59.9	0	625.4
1300	804.8	21.20	33.11	23.86	56.89	929.5	507.7	58.6	0	566.3
1400	864.8	36.20	26.58	38.23	63.42	875.1	391.5	55.1	0	446.6
1500	924.8	51.20	17.77	50.07	72.23	749.4	228.7	47.2	0	275.9
1600	984.8	66.20	7.43	59.84	82.57	386.0	49.9	24.3	0	74.2
1630	1014.8	73.70	1.86	64.14	88.14	12.5	0.41	0.79	0	1.20

Figure 2.11
Total solar irradiance for
Example 2.4.

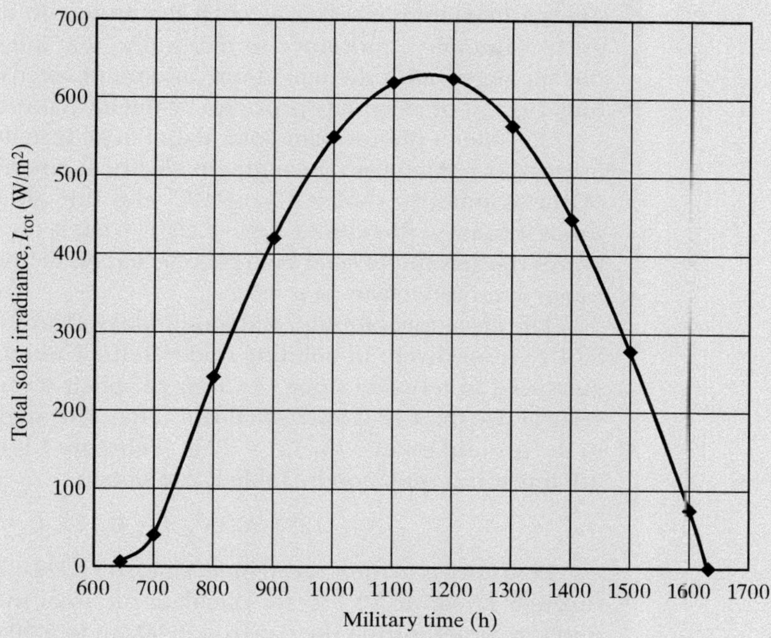

Thus, the total solar energy incident on the panel is

$$E_{solar} = H A_{collector}$$
$$= (13.78 \text{ MJ/m}^2)(8 \times 20) \text{ m}^2$$
$$= 2205 \text{ MJ}$$

Using Equation (2.17) the total electrical energy generated by the solar panel is

$$E_{elect} = \eta_{cell} E_{solar}$$
$$= (0.21)(2205 \text{ MJ})$$
$$= 463.1 \text{ MJ} \times \frac{1 \text{ kWh}}{3.6 \text{ MJ}}$$
$$= 128.6 \text{ kWh}$$

APPLICATION—OPTIMIZING SOLAR PANEL TILT ANGLE

Among other variables, the amount of solar energy incident upon a solar panel depends to a great extent on the tilt angle, β_2, of the panel. The solar panel of a photovoltaic system (particularly small commercial and residential) does not typically track the sun but is fixed in position on a roof top or other suitable location. For a given azimuth angle, α_2, is there a panel tilt angle that yields a maximum insolation, H? If the answer to this question is *yes*, it is of great economic importance to determine that angle, for a solar panel with this tilt angle yields the maximum amount of electrical power, thereby offsetting the cost of electricity provided by the local power utility.

Consider a photovoltaic solar panel to be installed on the roof of a home in Dearborn, Michigan, to augment electrical power from the grid. The pitch of the symmetric roof is "5 on 12," and one side of the roof is facing a south-southwest direction ($\alpha_2 = +22.5°$). What is the panel tilt angle that maximizes the amount of solar energy collected on February 1? Assume a standard foreground reflectivity of $\rho = 0.2$.

The elevation, latitude, and longitude of Dearborn are 180 m, 42.32°, and 83.18°, respectively. In building trades in the United States, roof pitch is often measured in terms of slope. A "5 on 12" pitch means that the roof has a 5-ft vertical rise for a 12-ft horizontal run. Thus, the angle of the roof with respect to the ground is $\tan^{-1}(5/12) = 22.6°$. February 1 is day 32 of the year, and we use linear interpolation in Table 2.2 to obtain

$$A = 1225 \text{ W/m}^2, B = 0.143, C = 0.059$$

Dearborn is in the eastern time zone, so $LSTM = 75°$. Using Equations (2.1) through Equation (2.14), we calculate the total solar irradiance, I_{tot}, for 11 different times during the day from 8:00 am to 5:30 pm for tilt angles of 22.6° (panel mounted flat on the roof), 30°, 45°, 60°, 70°, and 80°, making a total of 66 separate calculations for I_{tot}. (This reinforces the earlier recommendation to incorporate the computational procedure in a computer-based tool!) We then use Equation (2.15) to calculate the daily insolation, H, for each tilt angle.

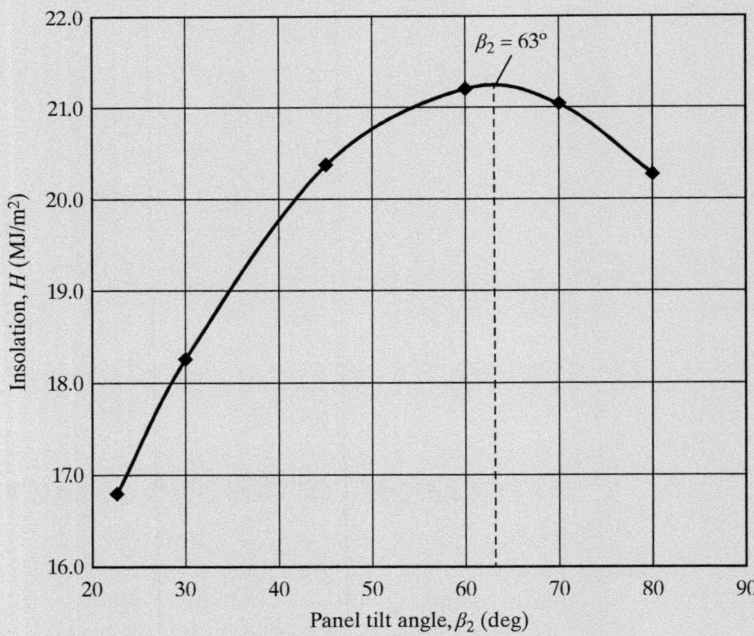

Figure 2.12 Optimum panel tilt angle.

A graph of insolation as a function of tilt angle reveals the *optimum* tilt angle. As shown in Figure 2.12, the optimum panel tilt angle is approximately 63°. Hence, the angle between the roof and panel is $(63 - 22.6)° = 40.4°$. It should be emphasized that this angle applies to one day only, February 1, and is not the optimum tilt angle for other days of the year. To find the optimum panel tilt angle for the entire year, insolation would have to be calculated for at least one day of each month (typically, the 21st day) for a range of panel tilt angles. The panel tilt angle with the highest annual insolation would be the optimum angle. This overall optimum tilt angle would be a compromise between winter and summer when the sun is at the lowest and highest positions, respectively, in the sky. It should also be emphasized that a solar panel is sometimes mounted flat on a roof even though the roof pitch is not the optimum angle. This is usually done because a solar panel mounted flat on the roof blends in with the roof line and is more visually appealing. Furthermore, local building codes or community standards may place restrictions on rooftop solar panel installations.

The optimum panel tilt angle for a *specific time* on a given day may be obtained by performing a maximization of total solar irradiance, I_{tot}, with respect to panel tilt angle, β_2. Differentiating Equation (2.9) with respect to β_2, setting the result to zero and solving for β_2, we obtain

$$\beta_{2,opt} = \tan^{-1}\left[\frac{2\cos\beta_1\cos(\alpha_1 - \alpha_2)}{C(1 - \rho) + (2 - \rho)\sin\beta_1}\right] \tag{2.18}$$

The derivation of Equation (2.18) is left as an exercise for the student in Problem 2.13.

The electrical output power of a photovoltaic system can be increased if the panel tracks the sun in two axes. The first axis is elevation, which corresponds to the tilt angle, β_2, and the second axis is azimuth, which corresponds to the azimuth angle, α_2. This approach is cost prohibitive in residential and small commercial photovoltaic systems because motors, sensors, and controls are required. To reduce cost, single-axis tracking (typically, elevation) is sometimes employed. Even higher output power can be achieved by using mirrors or lenses to concentrate the incident solar radiation onto the solar panels. This type of system, referred to as a *concentrating photovoltaic* system, requires fewer solar cells than a standard photovoltaic system but is feasible only for large-scale power production applications.

PRACTICE!

1. Find the total solar irradiance for a horizontal photovoltaic solar panel in Santa Fe, New Mexico, at 9:00 am, 1:00 pm, and 4:00 pm on October 21. Assume a foreground reflectivity of $\rho = 0.2$.

   Answer: 500 W/m^2, 697 W/m^2, 225 W/m^2

2. The solar panel in Practice Problem 1 has a surface area of 16 m^2. If the efficiency of the cells in the panel is 0.18, find the electrical energy generated by the panel on October 21.

   Answer: 14.2 kWh

3. Find the optimum tilt angle for a photovoltaic solar panel in Pittsburgh, Pennsylvania, on December 21. Also, find the insolation corresponding to this tilt angle. The panel faces due south, and the foreground is covered with snow.

Answer: $75°$, $22.6 \ \mathrm{MJ/m^2}$

PROFESSIONAL SUCCESS—THE AMERICAN SOLAR CHALLENGE

Buildings are not the only structures where photovoltaic solar panels can be used as a source of power. The American Solar Challenge (ASC), an event held every other summer, is an engineering competition in which college and university students from across the world design and build a solar-powered car and then compete with other schools in a multiday, 1200- to 1800-mile road race across North America. This event requires a year of intense work on the part of the student teams. This focused effort typically takes place during engineering students' senior year, with the ASC being the culmination of their senior capstone project. Each solar-powered car must be designed to meet strict performance specifications and race regulations, and successful cars are optimized and tested as much as possible before the race begins. Student teams submit vehicle design reports and other documentation to race officials for approval prior to the competition. Cars also undergo a thorough inspection to assure that they comply with all specifications and race rules. If the cars pass this hurdle, they must still qualify for the ASC by completing a predetermined number of laps in the Formula Sun Grand Prix (FSGP) track race (see Figure 2.13). On years when the ASC is held, the FSGP serves as a qualifying race for this competition. The winner of the ASC is determined by the total elapsed time to complete the race route.

Figure 2.13 Cars qualifying for the American Solar Challenge.
(David Parsons/NREL)

The ASC, and other projects like it, prepares engineering students for the transition from school to professional engineering practice. A year-long design project in the senior year integrates many of the principles and concepts learned throughout the students' engineering program. The ASC incorporates experimentation, problem formulation and solution, engineering mechanics, thermodynamics, dynamics, electrical circuits, instrumentation and controls, manufacturing, and, of course, design. This project also helps students learn the so-called soft skills of team work, communication, professionalism, and ethics.

2.4 SOLAR THERMAL SYSTEMS

In this section we consider two types of solar thermal systems. The first type is employed to heat water for domestic hot water use but is sometimes used to heat air for a living space or provide heat for other purposes. The second type of solar thermal system is designed to generate electrical power on a large scale. Fundamental engineering principles of these two systems and the procedures for doing basic energy analyses of them will now be described.

2.4.1 Domestic Water Heating

A typical system for heating domestic water was introduced earlier in the chapter and is illustrated in Figure 2.1. This figure shows only the primary components to illustrate the basic principle of operation. Components such as a means of energy storage, controls, valves, and other mechanical and electrical components are omitted for the sake of simplicity.

The collector of a domestic water heating system is called a *flat plate collector*, and its purpose is to absorb solar radiation, thereby heating a *working fluid* that is pumped through it. Working fluids in domestic water heating systems are usually an ethylene-glycol mixture (antifreeze). As illustrated in Figure 2.14, the collector consists of an enclosure with a bank of tubes under a glazing (glass or other type of transparent material). The fraction of the incident solar radiation that is

Figure 2.14
Flat plate solar collector.

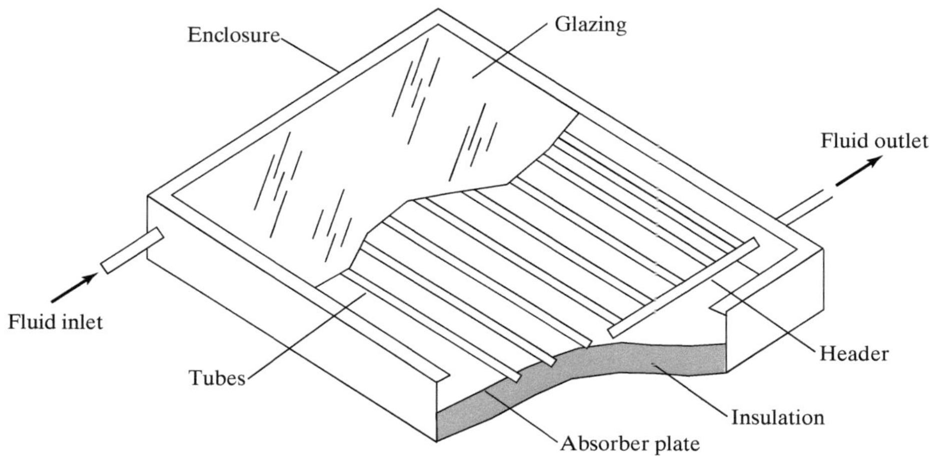

Figure 2.15
A roof-mounted flat plate solar collector for heating water.

(Pavel Vakhrushev/ Shutterstock)

transmitted through the glazing strikes the absorber plate and tubes, while the remainder of the incident radiation is absorbed by and reflected from the glazing. Part of the radiation that is absorbed by the glazing is reradiated and convected to the absorber plate and tubes, and the rest is reradiated and convected to the surroundings and is therefore lost energy. The fraction of incident radiation that is reflected from the glazing is also lost energy to the surroundings. In a well-designed collector, most of the radiation transmitted through the glazing is absorbed by the absorber plate and tube surfaces and subsequently transferred to the working fluid by conduction and convection. The absorber plate is backed with insulation to minimize heat conduction losses. Warm working fluid exits the collector and passes through a *heat exchanger* in a storage tank where the fluid's heat is transferred to the domestic water. (Refer to Figure 2.1.) Cool working fluid returns to the collector in a closed loop to be reheated by solar radiation. This type of system is normally found in residential and small commercial applications and does not have sun tracking capability. A roof-mounted flat plate collector is shown in Figure 2.15.

Solar energy absorbed by the working fluid is pumped from the collector to the heat exchanger in the storage tank. Because of energy losses in the collector, connecting pipes, and heat exchanger, only a portion of this absorbed energy is delivered to the domestic water. A thorough energy analysis of the system requires a knowledge of fluid mechanics and heat transfer, concepts that are typically covered in junior- or senior-level engineering courses. However, a simplified energy analysis can be used to demonstrate a few fundamental principles.

By applying the first law of thermodynamics (law of conservation of energy) to the working fluid, we can estimate the temperature of the working fluid as it enters the heat exchanger and thereby estimate the temperature of the domestic water in the storage tank. According to the first law of thermodynamics, the heat transfer to the working fluid in the collector, $\dot{Q}$, is related to the fluid's inlet and outlet temperatures, T_i and T_o, by the relation

$$\dot{Q} = \dot{m}c(T_o - T_i) \tag{2.19}$$

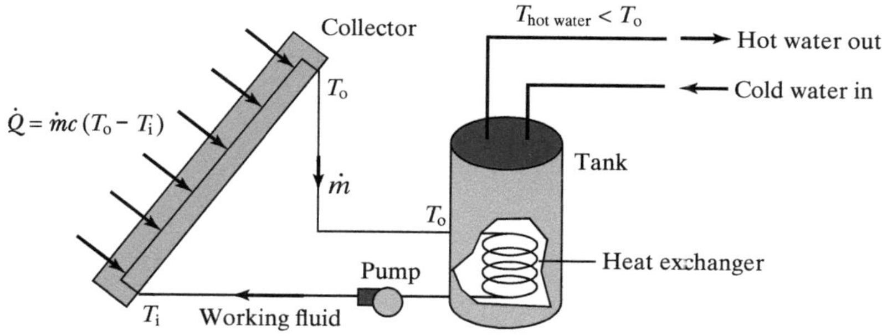

Figure 2.16
The domestic hot water temperature is limited by the outlet temperature of the working fluid.

where $\dot{m}$ and c are the mass flow rate and specific heat, respectively, of the working fluid. The quantities $\dot{m}$ and c have SI units of kg/s and J/kg·°C, respectively. Equation (2.19) applies to situations in which the heat transfer is steady and the fluid flows at a steady rate. The use of this equation in a domestic water heating system yields approximate results because the heat transfer varies considerably as the sun moves across the sky. Flow rate is steady if the pump runs continuously, but the flow cycles on and off when the system is thermostatically controlled to do so. This type of system typically runs intermittently and incorporates a means of energy storage so that domestic water can be heated during periods when the sun is not shining. Assuming no heat losses from the connecting pipes between the collector and heat exchanger, the temperature of the working fluid at the collector outlet, T_o, and the temperature of the working fluid at the heat exchanger inlet are equal. Clearly, the temperature of the domestic water in the storage tank cannot be any higher than T_o. In fact, the temperature of the water leaving the tank is lower than T_o due to heat transfer effects in the heat exchanger.

This simplified energy analysis approach, illustrated in Figure 2.16, is demonstrated in the following example.

EXAMPLE 2.5

A domestic water heating system in Chicago, Illinois, incorporates a roof-mounted flat plate solar collector with a tilt angle of 70°. The glazing surface measures 3.4 m × 5.6 m, and the collector faces due south. The working fluid is a mixture of ethylene glycol and water with a specific heat of $c = 3320$ J/kg·°C and a mass flow rate of $\dot{m} = 0.10$ kg/s. Assume that 60 percent of the total solar irradiance is absorbed by the working fluid and that $T_i = 22$°C (room temperature). What is the temperature of the working fluid at the outlet of the collector for conditions at 2:00 pm on January 21?

SOLUTION

Using the computational procedure discussed in Section 2.2.2, we find that the total solar irradiance for this collector is $I_{\text{tot}} = 935$ W/m². Multiplying this quantity by the surface area and 0.60, the "efficiency" of the collector, we obtain the heat transfer

$$\dot{Q} = 0.60\, I_{\text{tot}}\, A$$
$$= (0.60)\,(935\text{ W/m}^2)\,(3.4\text{ m} \times 5.6\text{ m})$$
$$= 1.068 \times 10^4 \text{ W}$$

(continued)

Using Equation (2.19), the temperature of the ethylene-glycol/water mixture at the outlet of the collector is

$$T_o = \frac{\dot{Q}}{\dot{m}c} + 22\overset{\circ}{}\mathrm{C}$$

$$= \frac{1.068 \times 10^4 \,\mathrm{W}}{(0.10 \,\mathrm{kg/s})\,(3320 \,\mathrm{J/kg \cdot {}^{\circ}C})} + 22\overset{\circ}{}\mathrm{C}$$

$$= 54.2\overset{\circ}{}\mathrm{C}\;(129.5\overset{\circ}{}\mathrm{F})$$

Due to heat transfer effects in the heat exchanger, the temperature of the domestic hot water would be a few degrees lower than this value, perhaps 45°C to 50°C (113°F to 122°F), a reasonable temperature range for domestic use.

The foregoing analysis provides only an estimate of the "instantaneous" domestic hot water temperature and does not address the amount of energy required or energy storage. For methods of conducting a more extensive engineering analysis of a solar domestic water heating system, the suggested readings at the end of the chapter should be consulted.

2.4.2 Concentrating Solar Power

Concentrating solar power (CSP) was introduced in Section 2.2.1. Unlike a photovoltaic system, which generates DC power, a CSP plant generates AC power. A concentrating power plant resembles a conventional power plant except that the source of energy is solar radiation rather than a fossil fuel. As illustrated in Figure 2.3, a working fluid that is heated by the sun in a receiver passes through a heat exchanger where the fluid transfers heat to water, converting the water to steam. A turbine converts the thermal energy of the steam to mechanical energy of a rotating shaft, and a generator converts the mechanical energy of the rotating shaft to electrical energy.

Because a CSP plant is basically a conventional power plant that uses solar radiation as a heat source, it is subject to the same laws of thermodynamics as a conventional power plant. A conventional power plant is fundamentally a **heat engine**, a device that converts thermal energy to work. **Thermal efficiency** of a heat engine, denoted η_{th}, is defined as the shaft work output of the turbine divided by the heat input for the system,

$$\eta_{th} = \frac{W_{out}}{Q_{in}} \tag{2.20}$$

where, in the case of CSP, the quantity Q_{in} is the heat absorbed by the working fluid in the receiver. The second law of thermodynamics states that it is impossible for a heat engine to convert all the heat it receives from a high-temperature source to work. Hence, the value of η_{th} is limited to values less than 1. A well-known theorem from thermodynamics states that for an *ideal* heat engine operating between source and sink temperatures of T_H and T_L, respectively, the maximum thermal efficiency is given by the relation

$$\eta_{th,\,ideal} = 1 - \frac{T_L}{T_H} \tag{2.21}$$

where T_H and T_L are expressed in absolute temperature units of kelvin (K) or rankine (R). Thermal efficiency given by Equation (2.21) is the maximum possible thermal efficiency a heat engine can have, and is referred to as the **Carnot efficiency**, in honor of the French engineer Sadi Carnot. In a CSP plant, T_L is approximately equal to the temperature of the surroundings, and T_H is the highest temperature in the system, the temperature of the working fluid in the receiver. It can be seen in Equation (2.21) that for a given value of T_L, ideal thermal efficiency increases with T_H. Hence, the maximum possible thermal efficiency of a CSP system is achieved by adding heat to the working fluid at the maximum possible temperature. This task is accomplished by concentrating sunlight. Depending on the type and design of the concentrator, working fluid temperatures may be 1500°C or higher.

CSP plants incorporate three main types of concentrators—heliostats, parabolic troughs and parabolic dishes (see Figure 2.2). Heliostats and parabolic dishes focus the sun's rays to a point, whereas parabolic troughs focus the sun's rays to multiple points (a line). A heliostat is a flat mirror that has dual-axis (elevation and azimuth) tracking, thereby maintaining a focus of reflected sunlight on the receiver at the top of the tower. A heliostat CSP system is shown in Figure 2.17. Likewise, a parabolic dish has dual-axis tracking. A parabolic trough, however, requires only single-axis (elevation) tracking because sunlight still focuses at a line at any azimuth angle. Only the direct component of solar irradiance is of importance for CSP systems since the mirrored surfaces are designed to track and concentrate direct sunlight. A parabolic trough CSP system is shown in Figure 2.18.

As a consequence of the mathematical properties of the parabola, sunlight reflects from the concentrator's mirrored surface and converges at the focus of the parabola, as illustrated in Figure 2.19. The parabolic mirror must face directly into the sun for this to occur, which is why tracking is required. Furthermore, a fundamental law of optics for a specular (mirror-like) surface states that the angles

Figure 2.17
The Gemasolar power plant, located in Seville, Spain, has a power-generation capacity of 19.9 MW.

(Greg Glatzmaier/NREL 19807 at Gemasolar Plant owned by Torresol Energy)

between a normal to the surface and the incident and reflected rays are equal. These angles are labeled φ in Figure 2.19. The aperture diameter and focal length are also indicated in the figure.

It is important to emphasize that the amount of solar energy that the receiver absorbs cannot be greater than the amount of solar energy that is incident on the concentrator. This is in accordance with the conservation of energy principle (first law of thermodynamics). For example, if the direct solar irradiance on the concentrator at a given time is 800 W/m^2 and the aperture area of the concentrator is 1 m^2, the receiver cannot absorb more solar power than 800 W/m$^2 \times 1$ m$^2 = 800$ W. How then does the working fluid in the receiver get so hot? The answer is that the concentration of sunlight increases the *heat flux* at the receiver and therefore the temperature of the working fluid. Heat flux is power per unit area and is expressed

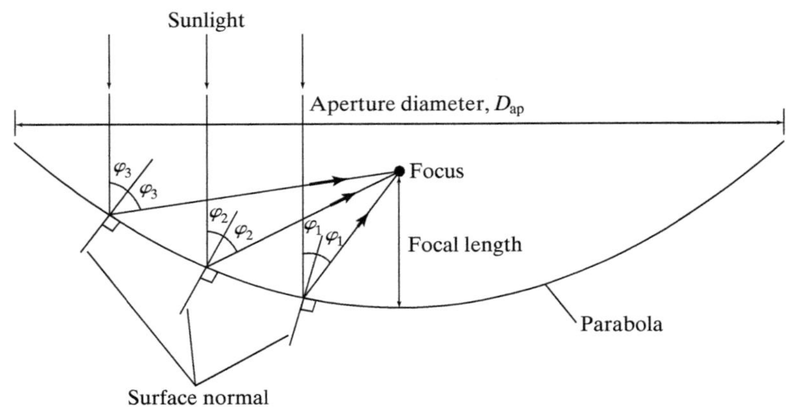

in units of W/m^2. The heat flux at the receiver can be approximated using the **concentration ratio**, defined by the relation

$$C = \frac{A_{\text{ap}}}{A_{rec}} \tag{2.22}$$

where A_{ap} is aperture area of the trough or dish (projected area of the concentrator) and A_{rec} is receiver area. Concentration ratios for trough and dish concentrators can be as high as approximately 200 and 4.5×10^4, respectively.

EXAMPLE 2.6

The direct solar irradiance on a dish concentrator is $750\ W/m^2$, and the dish aperture diameter is 2.6 m. The concentration ratio is 1200, the temperature of the working fluid in the receiver is 1000°C and the temperature of the surroundings is 20°C. Find the solar power incident on the concentrator, the heat flux at the receiver and the Carnot efficiency.

SOLUTION

The solar power incident on the concentrator, $\dot{Q}_{\text{con}}$, is the product of direct solar irradiance and aperture area,

$$\dot{Q}_{\text{con}} = I_D A_{\text{ap}} = I_D \pi D_{\text{ap}}^2/4$$
$$= (750\ W/m^2)\pi(2.6\ m)^2/4$$
$$= 3982\ W$$

Assuming no optical or thermal losses in the concentrator, the heat flux at the receiver is the product of the direct solar irradiance on the concentrator and the concentration ratio,

$$\dot{Q}''_{\text{rec}} = I_D C$$
$$= (750\ W/m^2)(1200)$$
$$= 9.00 \times 10^5\ W/m^2$$

From Equation (2.21), the Carnot efficiency is

$$\eta_{\text{th, ideal}} = 1 - \frac{T_L}{T_H}$$
$$= 1 - \frac{(20 + 273)\ K}{(1000 + 273)\ K}$$
$$= 0.77$$

Hence, the actual thermal efficiency of the CSP plant that uses this dish cannot exceed 77 percent. Thermal efficiency is one of four efficiencies that apply to concentrating solar power systems, as explained in the next section.

In a concentrating solar power system, there are four energy conversion efficiencies that must be considered. The first is thermal efficiency, η_{th}, which has already been discussed. The other three efficiencies apply to specific components of CSP systems: the concentrator, receiver, and electrical generator. The concentrator is not optically perfect and therefore does not reflect all of its incident solar radiation onto the receiver. A small amount of the incident solar radiation is absorbed by the mirrored surfaces where it is transferred by conduction through the structure and subsequently lost to the surroundings by convection and radiation. A fraction of the solar energy reflected from the concentrator onto the receiver is not absorbed by the working fluid but is transferred as heat to the receiver structure and then lost to the surroundings by convection and radiation. Finally, a fraction of the mechanical energy produced by the turbine is not converted to electrical energy due to various energy loss mechanisms in the generator.

Combining these four efficiencies, the overall efficiency of the concentrating solar power system is expressed as

$$\eta_{\text{overall}} = \eta_{\text{th}}\eta_{\text{con}}\eta_{\text{rec}}\eta_{\text{gen}} \tag{2.23}$$

where η_{con}, η_{rec}, and η_{gen} are the efficiencies of the concentrator, receiver, and generator, respectively. Solar reflectivity of the concentrator's mirrored surfaces is high, so η_{con} is typically greater than 0.90. The efficiency of a well-designed receiver, η_{rec}, ranges from approximately 0.75 to 0.85, and the efficiency of large electrical generators can be as high as 0.97. Thermal efficiency, η_{th}, cannot exceed the Carnot efficiency, $\eta_{\text{th,ideal}}$, and is a function of the thermal and mechanical performance of the heat exchanger, condenser, pump, turbine, and connecting pipes.

An instantaneous overall efficiency (efficiency at a moment of time) of a CSP plant can be expressed as the electrical output power divided by the solar input power,

$$\eta_{\text{overall}} = \frac{P_{\text{elect}}}{\dot{Q}_{\text{con}}} \tag{2.24}$$

where P_{elect} is output power of the electrical generator and $\dot{Q}_{\text{con}}$ is solar power that is incident on the concentrator. An average overall efficiency can be calculated using Equation (2.23) if time-averaged values of P_{elect} and $\dot{Q}_{\text{con}}$ are used. Overall efficiency can also be expressed as

$$\eta_{\text{overall}} = \frac{E_{\text{elect}}}{E_{\text{solar}}} \tag{2.25}$$

where the quantities E_{elect} and E_{solar} are the electrical output energy and solar input energy, respectively, for a prescribed period of time.

EXAMPLE 2.7

Consider a parabolic trough concentrator with an aperture width and length of 2 m and 30 m, respectively. During a certain day, the average direct solar irradiance on the concentrator is 650 W/m². The efficiencies of the concentrator, receiver, and electrical generator are 0.92, 0.80, and 0.96, respectively. If the temperature of the working fluid in the receiver is 300°C, find the maximum possible electrical output power. The temperature of the surroundings is 10°C.

SOLUTION

The average solar power that is incident on the concentrator is

$$\dot{Q}_{con} = I_D A_{ap}$$
$$= (650 \text{ W/m}^2)\,(2 \text{ m} \times 30 \text{ m})$$
$$= 3.90 \times 10^4 \text{ W}$$

From Equation (2.21), the Carnot efficiency is

$$\eta_{th,\,ideal} = 1 - \frac{T_L}{T_H}$$
$$= 1 - \frac{(10 + 273) \text{ K}}{(300 + 273) \text{ K}}$$
$$= 0.51$$

Using Equation (2.23), the maximum (ideal) overall efficiency is

$$\eta_{overall,\,max} = \eta_{th,\,ideal}\,\eta_{con}\,\eta_{rec}\,\eta_{gen}$$
$$= (0.51)\,(0.92)\,(0.80)\,(0.96)$$
$$= 0.36$$

Applying Equation (2.24) to the maximum electrical output power, we obtain

$$P_{elect,\,max} = \eta_{overall,\,max}\,\dot{Q}_{con}$$
$$= (0.36)\,(3.90 \times 10^4 \text{ W})$$
$$= 1.40 \times 10^4 \text{ W} = 14.0 \text{ kW}$$

This is the maximum possible electrical output power because the overall efficiency is based on $\eta_{th,ideal}$, not the *actual* thermal efficiency, η_{th}. For all heat engines, $\eta_{th} < \eta_{th,ideal}$, so the actual electrical output power is less than 14.0 kW.

EXAMPLE 2.8

A CSP plant is being designed to augment electrical power from the main grid to a small residential community. The plant is to consist of an array of parabolic dish concentrators with an aperture diameter of 2.8 m. The efficiencies of the system are estimated to be

$$\eta_{th} = 0.38, \; \eta_{con} = 0.95, \; \eta_{rec} = 0.82, \; \eta_{gen} = 0.97$$

As a starting point in the design, an engineer uses 850 W/m^2 as a direct solar irradiance averaged over a 12-hour period during which the sun shines on a typical day. The average electrical power consumption of a home in the United States is approximately 30 kWh per day. If the community has 1600 homes, how many dish concentrators are required to provide 20 percent of the power? Also, estimate the amount of land required for the concentrators.

(continued)

SOLUTION

We first calculate the amount of solar energy incident on one dish concentrator during a time period of $\Delta t = 12$ h. Recognizing that $1\text{ W} = 1\text{ J/s}$, we have

$$
\begin{aligned}
E_{\text{solar}} &= I_D \Delta t A_{\text{ap}} = I_D \Delta t \pi D_{\text{ap}}^2/4 \\
&= (850\text{ J/s·m}^2)(12\text{ h} \times 3600\text{ s/h})\pi(2.8\text{ m})^2/4 \\
&= 2.26 \times 10^8\text{ J} = 226\text{ MJ}
\end{aligned}
$$

Using Equation (2.23), the overall efficiency of the concentrating solar power plant is

$$
\begin{aligned}
\eta_{\text{overall}} &= \eta_{\text{th}}\,\eta_{\text{con}}\,\eta_{\text{rec}}\,\eta_{\text{gen}} \\
&= (0.38)(0.95)(0.82)(0.97) \\
&= 0.29
\end{aligned}
$$

Using Equation (2.25), the electrical output energy per day for one concentrator is

$$
\begin{aligned}
E_{\text{elect}} &= \eta_{\text{overall}} E_{\text{solar}} \\
&= (0.29)(226\text{ MJ}) \\
&= 65.5\text{ MJ} \times (1\text{ kWh}/3.6\text{ MJ}) \\
&= 18.2\text{ kWh (one concentrator)}
\end{aligned}
$$

The electrical energy that the CSP plant must supply per day is

$$
\begin{aligned}
E_{\text{elect}} &= (0.20)(1600\text{ home})(30\text{ kWh/home}) \\
&= 9.60 \times 10^3\text{ kWh (all concentrators)}
\end{aligned}
$$

Thus, the number of dish concentrators, N, required to meet the power requirement of the community is

$$
\begin{aligned}
&= 9.60 \times 10^3\text{ kWh}/(18.2\text{ kWh}) \\
&= 527
\end{aligned}
$$

One possible arrangement is a 23×23 square array of concentrators. The concentrators require a space around them to assure that they do not shade or interfere with adjacent concentrators and to allow space for maintenance. Estimating a required land area for each concentrator of three times the aperture area of a single dish (6.16 m^2), 527 concentrators take up $3 \times 527 \times 6.16\text{ m}^2 = 9739\text{ m}^2$ (2.4 acre) of land. Additional land area would be required for the other components of the power plant.

In the foregoing material on CSP, we have not addressed energy storage. A CSP plant has the capability of integrating thermal energy storage (TES) such that when the system generates more energy than is needed to supply the instantaneous electricity demand, that energy is stored for future use, as illustrated in Figure 2.20. Solar energy is stored by circulating the working fluid through a separate heat exchanger that transfers the energy to a secondary fluid in a storage tank. During daylight hours, excess solar energy is delivered to the stored fluid to be used later in the day. After sunset and during the early evening hours when electricity demand

Figure 2.20
Storage and extraction of solar thermal energy.

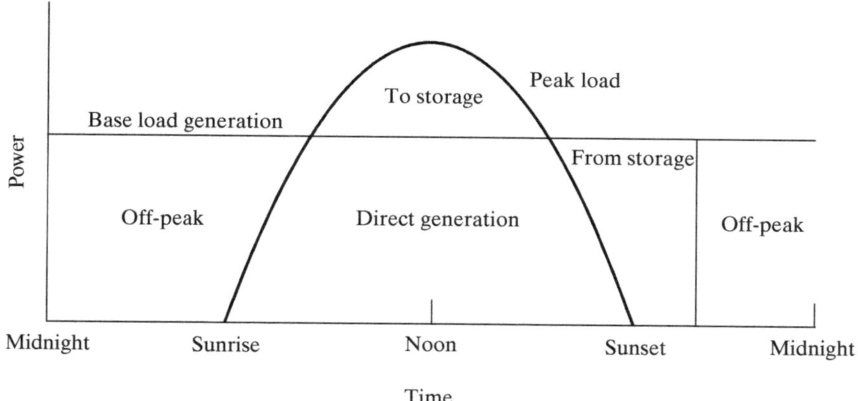

is typically the highest (peak load), the stored solar energy is extracted from the fluid in the tank. For additional information on thermal energy storage systems for concentrating solar power plants and methods for analyzing them, consult the suggested readings at the end of the chapter.

PRACTICE!

1. A gravel roof-mounted, south-facing flat plate solar collector with a glazing surface measuring 2.5 m × 5.5 m has a tilt angle of 60°. The working fluid is a mixture of ethylene glycol and water with a specific heat of $c = 3320$ J/kg·°C and a mass flow rate of 0.075 kg/s. Assuming that 55 percent of the total solar irradiance is absorbed by the working fluid and that $T_i = 20$°C, find the temperature of the working fluid at the outlet of the collector for conditions at 12:00 noon on February 21. The collector is located in Albany, New York, longitude 73.76°, latitude 42.65° and elevation 0 m (Albany borders the Hudson River, a sea-level river).

    Answer: 50.7°C

2. The aperture diameter of a dish concentrator is 1.9 m, and the concentration ratio is 1000. The temperature of the working fluid in the receiver is 1250°C, and the temperature of the surroundings is 25°C. If the direct solar irradiance on the dish is 680 W/m², find the solar power incident on the concentrator, the heat flux at the receiver, and the Carnot efficiency.

    Answer: 1928 W, 6.80×10^5 W/m², 0.80

3. A parabolic trough concentrator has an aperture width and length of 1.6 m and 24 m, respectively. The average direct solar irradiance on the concentrator on a given day is 590 W/m². The efficiencies of the concentrator, receiver, and electrical generator are 0.90, 0.82, and 0.94, respectively. Find the maximum possible electrical output power for a working fluid temperature at the receiver of 400°C. The temperature of the surroundings is 15°C.

    Answer: 8991 W

2.5 ENVIRONMENTAL CONSIDERATIONS

Solar energy is considered to be one of the most environmentally friendly types of renewable energy. During operation, neither photovoltaic nor solar thermal systems introduce harmful emissions into the atmosphere. However, solar energy is not without environmental impacts.

Because the amount of solar energy that can be intercepted by a collector is limited, both photovoltaic and solar thermal systems require a significant amount of space. On a commercial scale, a large area of land is required, and on a residential scale, the space on a roof top limits the amount of energy that can be gathered. Furthermore, a roof top installation must share space with other things such as vent pipes, sky lights, and other mechanical and electrical systems. However, solar thermal systems generate an insignificant amount of noise, and photovoltaic systems generate none.

The main component of most photovoltaic modules is silicon, a semiconductor material that is very energy intensive to manufacture. A significant amount of electrical energy is required to produce silicon, and because the majority of electricity is still generated by fossil-fuel power plants, the production of silicon introduces harmful substances into the atmosphere. In addition, the manufacture of silicon involves toxic chemicals such as silane gas, sulfur hexafluoride, and cadmium. It is important to note that the manufacture of other electronic devices such as televisions and computers involve many of the same toxic materials, so the environmental impacts are not limited to photovoltaic modules.

Photovoltaic panels pose a potential risk of electrical shock, but as with other electrical power systems, sound engineering design minimizes this hazard. Photovoltaic panels have a visual impact on the environment, particularly in residential installations. To minimize aesthetically unattractive features, architects, and designers strive to blend photovoltaic systems with existing architectural elements such as roof lines.

In large commercial installations, space is required around each collector or panel to prevent overshadowing. Potential uses of this space include growing crops or, if the site is suitable, generating additional renewable energy from wind turbines.

In concentrating solar power plants, birds have been killed by attempting to fly across the concentrated beam focused on the heliostat. As birds fly into the beam in front of the heliostat, they are scorched in midair by the intense heat produced by the concentrated solar energy.

SUMMARY

Solar energy is the most vital of all energy sources because no life on earth could exist without it. With the exception of nuclear and geothermal energy, all forms of energy directly or indirectly depend on the sun. At the outer edge of the atmosphere, the solar heat flux is 1366 W/m^2, a value known as the solar constant.

There are two basic types of solar energy systems: solar thermal and photovoltaic. On a small scale, the first type of system is used to heat water for domestic use or air in a living space. On a large scale, it is used to generate electrical energy. The second type of energy system converts sunlight directly into electrical energy using solar cells.

All solar energy systems employ either a collector or concentrator. Knowing the latitude, longitude, day of the year, time of day, surface orientation, and atmospheric conditions, a set of equations can be used to calculate the amount of solar irradiance on the surface. Total solar irradiance is the sum of three components: direct, scattered, and reflected irradiance. The main purpose of doing irradiance calculations is to maximize the amount of solar radiation collected and thus the amount of thermal or electrical energy that can be generated.

Solar insolation is the incident solar energy per unit surface area. Insolation is found by integrating the total solar irradiance over a specified time period, which is typically one day. The tilt angle of solar panel of a photovoltaic system can be optimized so as to maximize insolation at a given time or over a given time period. The typical range of solar cell efficiency is 15 to 25 percent.

The temperature of a domestic water heating system is limited by the temperature of the water at the exit of the collector, which can be calculated using the law of conservation of energy.

A concentrating solar power system is basically a conventional power plant that uses solar radiation as the heat source, thereby making it a type of heat engine. The efficiency of a concentrating solar power plant is limited by the Carnot efficiency. Concentrating solar power plants incorporate three types of concentrators: heliostats, parabolic troughs, and parabolic dishes. Working fluids of a concentrating power plant can be heated to temperatures as high as 1500°C by increasing the heat flux at the receiver. Four efficiencies are associated with concentrating solar power plants: thermal efficiency, concentrator efficiency, receiver efficiency, and generator efficiency. The overall efficiency is the product of these four quantities.

Solar energy is considered to be one of the most environmentally friendly types of renewable energy. However, solar energy systems impact the environment because collectors occupy a great deal of space. Furthermore, most photovoltaic solar panels are made from silicon, a very energy extensive material to manufacture, and the manufacture of silicon involves toxic chemicals which must be prevented from entering the natural environment. Lastly, solar collectors may be visually detracting.

KEY TERMS

Carnot efficiency	efficiency	solar energy
concentrating solar power	heat engine	solar thermal
concentration ratio	insolation	thermal efficiency
	photovoltaic	total solar irradiance

SUGGESTED READING

ASHRAE Handbook–HVAC Applications, American Society of Heating, Refrigerating and Air-conditioning Engineers, New York, 2011.

BOYLE, G., Ed., *Renewable Energy–Power for a Sustainable Future* 3rd Ed, Oxford, Oxford University Press, 2012.

BREEZE, P., *Power Generation Technologies*, Burlington, MA: Elsevier, 2005.

HAGEN, K.D., *Introduction to Engineering Analysis* 4th Ed, Upper Saddle River, NJ: Prentice Hall, 2014.

ZOOBA, A.F. and R.C. BANSAL, Eds., *Handbook of Renewable Energy Technology*, Hackensack, NJ: World Scientific, 2011.

PROBLEMS

For the following problems, it is recommended that you use the general analysis procedure of (1) problem statement, (2) diagram, (3) assumptions, (4) governing equations, (5) calculations, (6) solution check, and (7) discussion. This procedure is covered in Chapter 3 of Hagen, K.D., *Introduction to Engineering Analysis* 4th Ed.

The Solar Energy Source

2.1 For Denver, Colorado, on December 19 at 10:00 am, find the declination angle and apparent solar time.

2.2 For Atlanta, Georgia, on March 2 at 2:30 pm, find the declination angle and apparent solar time.

2.3 For Albany, New York, on August 20 at 11:00 am, find the declination angle and apparent solar time.

2.4 A collector, located in Columbus, Ohio, has a tilt angle of 60° and faces due south. If the foreground is concrete, find the total solar irradiance on the collector at 3:00 pm on January 21.

2.5 Consider a south-facing collector in Springfield, Illinois. If the collector tilt angle is 70°, find the total solar irradiance on the collector at 10:30 am on November 21. The foreground is a gravel roof.

2.6 A horizontal collector on a roof top is located in Austin, Texas. Find the total solar irradiance on the collector at 11:30 am on July 12.

2.7 The collector in Problem 2.4 has a surface area of 24 m^2. Find the total solar power incident on the collector at 1:00 pm. Also, find the insolation and the total solar energy incident on the collector during the day.

2.8 The collector in Problem 2.5 has a surface area of 18 m^2. Find the total solar power incident on the collector at 11:00 am. Also, find the insolation and the total solar energy incident on the collector during the day.

2.9 The collector in Problem 2.6 has a surface area of 60 m^2. Find the total solar power incident on the collector at 12:00 noon. Also, find the insolation and the total solar energy incident on the collector during the day.

Photovoltaic Systems

2.10 A manufacturing plant in San Jose, California, has a roof-mounted photovoltaic solar panel that measures 16 m $\times$ 30 m. The panel is horizontal and has an efficiency of 0.18. How much electrical energy does the panel generate on December 21?

2.11 A residential solar system in Montgomery, Alabama, incorporates a photovoltaic solar panel with a tilt angle of 60°. The roof-mounted panel, which measures 20 m^2, faces due south and has an efficiency of 0.16. How much electrical energy does the panel generate on October 21?

2.12 A classroom building at a university in Raleigh, North Carolina, has an array of photovoltaic solar panels on its roof for augmenting electrical power from the grid. The panels face due south and have a tilt angle and efficiency of 60° and 0.22, respectively. How much electrical energy do the panels generate on December 21 for a total panel surface area of 600 m^2?

2.13 Derive Equation (2.18).

2.14 Using Equation (2.18), calculate the optimum tilt angle for a solar panel in Salt Lake City, Utah, at 1:00 pm on January 21. The panel faces a south-south-west direction, and the foreground is covered with snow.

2.15 A photovoltaic solar panel is to be installed on the roof of a home in Phoenix, Arizona. The roof has a 4 on 12 pitch, and one side of the roof faces south-southeast. For November 21, find the optimum angle of the solar panel with respect to the roof.

2.16 An office complex in San Antonio, Texas, has a roof-mounted photovoltaic solar panel that measures 24 m × 36 m. The panel is horizontal and has an efficiency of 0.22. How much electrical energy does the panel generate on July 21?

2.17 A home owner in Sacramento, California, wishes to install photovoltaic solar panels on his south-facing roof. The pitch and reflectivity of the roof are 12 on 12 and 0.2, respectively. If the efficiency of the panels is 0.18, how much electrical energy can the home owner expect to generate on December 31 if he mounts the panels flat on the roof? The total surface area of the panels is 70 m^2.

2.18 Work Problem 2.17 for the optimum panel tilt angle for December 31.

2.19 After cost incentives totaling $6500 by state and local governments, the initial cost of a photovoltaic system for a residence is $11,000. If the annual energy production of the solar panels is 5750 kWh/y, and the energy cost is $0.10/kWh, what is the simple payback?

2.20 Design a photovoltaic system for a residence in Tempe, Arizona. The south-facing roof has a surface area of 45 m^2 and a 5 on 12 pitch. State all your assumptions.

2.21 A rover for gathering geological data on Mars has a photovoltaic solar panel that supplies electrical power to its sensors, motors, and other electrical components. The panel faces directly into the sun, measures 2.6 m × 1.2 m, and has an efficiency of 0.18. If the solar constant for Mars is 585 W/m^2, what is the electrical power capacity of this system?

Solar Thermal Systems

2.22 In a domestic water heating system, an ethylene-glycol/water mixture flows through the collector at a mass flow rate of 0.065 kg/s. The glazing surface measures 3 m × 5 m, and 550 W/m^2 of solar power is absorbed by the fluid mixture. If the inlet temperature of the fluid mixture is 25°C, what is the outlet temperature?

2.23 A domestic water heating system in Los Angeles, California, incorporates a roof-mounted flat plate solar collector with a tilt angle of 60°. The collector faces south-southeast, and the glazing surface measures 4.0 m × 6.2 m. The working fluid, which has a specific heat of $c = 3150$ J/kg·°C, enters the collector at a temperature of 21°C and flows through the collector at a mass flow rate of 0.10 kg/s. If 50 percent of the total solar irradiance is absorbed by the working fluid, find the temperature of the working fluid at the outlet of the collector at 12:00 noon on April 21.

2.24 The direct solar irradiance on a dish concentrator is 600 W/m^2, and the dish aperture diameter is 1.80 m. The concentration ratio is 2500, the temperature of the working fluid in the receiver is 950°C, and the temperature of the surroundings is 25°C. Find the solar power incident on the concentrator, the heat flux at the receiver, and the Carnot efficiency.

2.25 Consider a CSP plant with an array of 400 dish concentrators in Santa Fe, New Mexico. The concentrators, which have an aperture diameter of 2.75 m, are installed in a gravel field and have dual-axis tracking. For 12:00 noon on January 21, find the total solar power incident on the concentrators and the heat flux at the receivers if the concentration ratio is 3200.

2.26 What is the maximum possible thermal efficiency of a concentrating solar power plant if the temperature of the working fluid at the receivers is 1400°C and the ambient air temperature is 20°C?

2.27 Consider a parabolic trough concentrator with an aperture width and length of 1.4 m and 26 m, respectively. During a given day, the average direct solar irradiance on the concentrator is 650 W/m^2. The efficiencies of the concentrator, receiver, and electrical generator are 0.91, 0.83, and 0.95, respectively. If the temperature of the working fluid in the receiver is 600°C, find the maximum possible electrical output power. The temperature of the surroundings is 10°C.

2.28 A concentrating solar power plant is being designed to augment electrical power from the main grid to a small residential community. The plant is to consist of an array of parabolic dish concentrators with an aperture diameter of 2.2 m. The efficiencies of the system are estimated to be:

$$\eta_{th} = 0.30, \ \eta_{con} = 0.90, \ \eta_{rec} = 0.82, \ \eta_{gen} = 0.95$$

As a starting point in the design, an engineer uses 800 W/m^2 as a direct solar irradiance averaged over a 12-hour period during which the sun shines on a typical day. The average electrical power consumption of a home in the United States is approximately 30 kWh per day. If the community has 1200 homes, how many dish concentrators are required to provide 25 percent of the power? Also, estimate the amount of land required for the concentrators.

2.29 A concentrating solar power plant with a power-generation capacity of 1.75 MW costs $2.50 million to install. The fixed charge rate is 7.5 percent, the annual operation and maintenance cost is assumed to be 1.0 percent of the initial cost, and the plant capacity factor is 0.20. If the levelized replacement cost is averaged over an expected 25-year lifetime, what is the cost of energy?

Environmental Considerations

2.30 Write an essay on bird deaths resulting from birds' flying into the beams of concentrating solar power plants.

2.31 Write an essay on the environmental impacts of manufacturing silicon photovoltaic solar panels.

3 Wind Energy

Objectives

After reading this chapter, you will have learned:
- What wind energy is
- The capacity of wind for supplying global electrical power
- How wind speeds vary in time and space
- How much power is available in the wind
- The basic characteristics of the two main types of wind turbines
- How to calculate the electrical output power of wind turbines
- The basic environmental issues with wind energy

3.1 INTRODUCTION

Wind is one of the oldest renewable energy sources and was one of the first nonanimal sources of energy to be used by humans. Sailboats have been propelled by the wind for thousands of years, and windmills were traditionally used for pumping water, grinding grain and spices, sawing wood, and making paper materials and paints and dyes. The earliest documented windmills in western Europe date from the twelfth century A.D., although windmills were in use in Asia and other parts of the ancient world centuries earlier. By the eighteenth century A.D., the widely recognized multisail Dutch windmills were in widespread use throughout Europe. A traditional windmill is shown in Figure 3.1. Along with solar and other renewable energy sources, wind energy has been widely developed in recent years. The first wind turbine for generating electrical power was developed in the late nineteenth century. Wind energy technologies developed gradually during the early to mid-twentieth century, but during the oil crisis of the 1970s, wind energy research and development expanded rapidly. By the end of the twentieth century, wind energy played an integral role in

Figure 3.1
A traditional wind mill.

*(Nick Biemans/
Shutterstock)*

renewable energy development. Wind energy has several advantages such as no pollution, short implementation time, and relatively low capital cost. Currently, wind supplies approximately 1 percent of the world's electrical energy demand, but it is anticipated that wind energy will supply about 12 percent of the world's electrical energy demand by 2020.

Over the earth's landmasses, winds generate approximately 6.1×10^{21} J of energy annually. Dividing this energy by the number of seconds in a year, we obtain 1.9×10^{14} W, the global power generated by winds. The global power consumption in 2010 has been estimated to be 1.7×10^{13} W. Furthermore, the solar power at the earth's surface is 8.1×10^{16} W, approximately two orders of magnitude higher than the energy generated by winds. This comparison is somewhat misleading, however, because the quantity given for wind energy generation applies to landmasses only, and about 75 percent of the earth's surface is covered by oceans. Nevertheless, we can see that wind, like solar, has the capacity to supply the world's energy demands.

The objective of wind power systems is to convert the kinetic energy of moving air to electrical energy. This is accomplished using a special machine, commonly referred to as a **wind turbine**, that facilitates this energy conversion. A wind turbine converts kinetic energy of moving air to mechanical energy of a rotating shaft, which is then converted to electrical energy.

In this chapter we examine the source of wind energy and explain how much energy is available in the wind for conversion to electrical energy. We discuss the main types of wind turbines and explain their fundamental operating principles. We conclude the chapter by discussing the environmental impacts of wind energy.

3.2 THE WIND ENERGY SOURCE

Wind is the movement of atmospheric air from regions of high pressure to regions of low pressure. The main cause of winds is the uneven heating of the earth's surface by the sun, which depends on latitude, time of day, and the distribution of land and large bodies of water, particularly the oceans. Another cause of winds is fluid friction between the atmosphere and earth's surface, which allows the earth to drag the atmosphere around, thereby producing turbulence. Horizontal components of wind velocities are normally much greater than the vertical velocity components, which is why we normally think of wind as the horizontal movement of air.

3.2.1 Wind Speed Variations

Winds vary significantly in both space and time. The scales over which spatial and temporal variations occur range from centimeters to thousands of kilometers and seconds to years, respectively. Spatial variations are a function of terrain, ground cover, proximity to bodies of water, and other parameters. Shown in Figure 3.2 is a "topographical" wind speed map of the United States from the National Renewable Energy Laboratory (NREL). The map shows seven zones of wind speeds ranging from 0 m/s to about 12 m/s averaged over a one-year period. Similar maps are available from NREL for each of the four seasons of the year.

Temporal variations can be generally categorized as interannual, annual, diurnal, and short term. Interannual wind speed variations occur over time periods greater than one year. According to meteorologists, it takes at least five years of wind data to reliably determine average wind speeds at a given location. Annual

Figure 3.2
Wind speed and power map for the United States.

(Courtesy of the Pacific Northwest National Laboratory, operated by Battelle for the U.S. Department of Energy.)

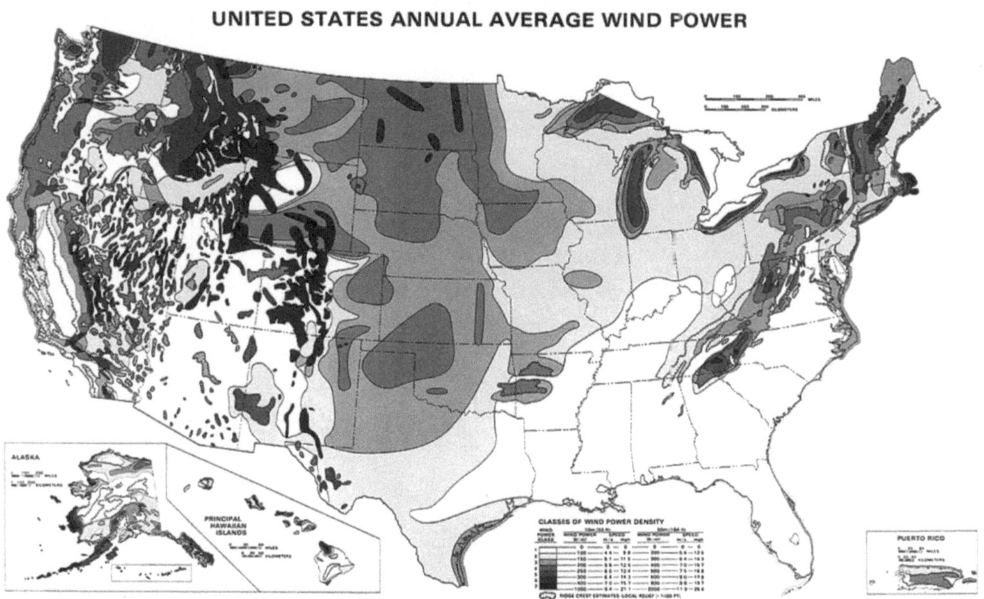

For a full-size, color version of the United States Annual Average Wind Power Map, please visit: http://rredc.nrel.gov/wind/pubs/atlas/maps/chap2/2-01m.html.

wind speed variations occur over time periods of seasons or months. For example, wind speeds tend to be maximum during the winter in mountainous regions of the United States, whereas wind speeds tend to be maximum during the spring in the Great Plains region. Diurnal wind speed variations occur on a daily time scale and are caused primarily by differential solar heating of the earth's surface. Wind speeds typically increase during the day and decrease during the hours from midnight to sunrise. Short-term wind speed variations refer to turbulence and wind gusts. Turbulence refers to wind speed fluctuations at random directions superimposed on the main wind speed. Because of the random nature of this phenomenon, turbulence is mathematically modeled using statistics. A wind gust is a sudden increase in wind speed in the direction of the main flow or a random direction. Good engineering design of wind power systems takes into account both spatial and temporal wind speed variations.

In addition to spatial and temporal wind speed variations, we also have an elevation (distance from the ground) variation. This wind speed variation is a consequence of the **boundary layer**, the region near a surface in which the velocity of a fluid varies from zero to the so-called free stream value. All fluids (liquids and gases) have *viscosity*, a property characterized by the fluid's resistance to flow due to friction. For instance, pancake syrup has a very high viscosity, whereas gases, such as air, have a very low viscosity. As a fluid flows across a surface, viscosity causes the fluid's particles to stick to the surface, yielding a fluid velocity of zero. Fluid velocity gradually increases with distance from the surface until the velocity reaches a constant value known as the free stream velocity. This defines the *velocity profile* of the boundary layer. Boundary layers occur in all situations where a fluid flows across a surface, including the flow of air across the surface of the earth. The boundary layer concept is illustrated in Figure 3.3.

Because wind speed increases with height, turbines are typically mounted on the top of tall towers. Wind speed at a given height above the ground can be estimated using the relation

$$\left(\frac{v}{v_0}\right) = \left(\frac{H}{H_0}\right)^{\alpha} \tag{3.1}$$

where v is wind speed at height H, v_0 is wind speed at height H_0, and α is the friction coefficient, a quantity that represents the "smoothness" of the terrain. Typical values of the friction coefficient for various types of terrain are given in Table 3.1.

Figure 3.3

A boundary layer is a region near a surface where the velocity of a fluid varies from zero to the free stream value.

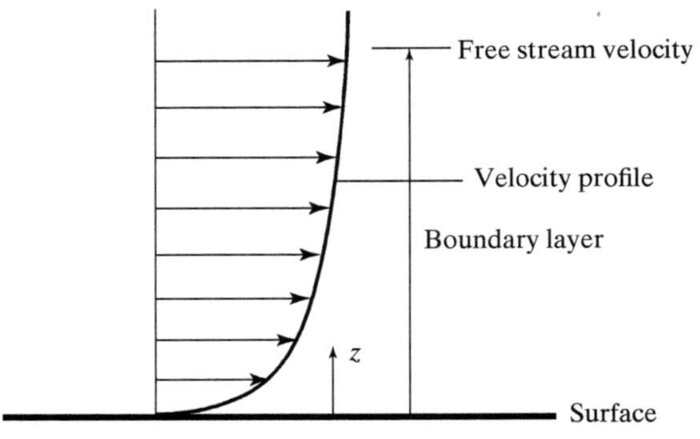

Table 3.1 Friction Coefficient for Types of Terrain

Type of terrain	Friction coefficient, α
Smooth hard ground, smooth water	0.10
Tall grass	0.15
High crops, hedges	0.20
Wooded ground with many trees	0.25
Small town with trees	0.30
Large city with tall buildings	0.40

EXAMPLE 3.1

The wind speed for a fully grown corn field at a height of 2 m from the ground is 4.0 m/s. Estimate the wind speed at heights of 30 m and 50 m.

SOLUTION

From Table 3.1, the friction coefficient for high crops is $\alpha = 0.20$. From Equation (3.1), the wind speed at a height of 30 m is

$$
\begin{aligned}
v &= (H/H_0)^{\alpha}(v_0) \\
&= (30 \text{ m}/2 \text{ m})^{0.20}(4.0 \text{ m/s}) \\
&= 6.9 \text{ m/s}
\end{aligned}
$$

and the wind speed at a height of 50 m is

$$
\begin{aligned}
&= (50 \text{ m}/2 \text{ m})^{0.20}(4.0 \text{ m/s}) \\
&= 7.6 \text{ m/s}
\end{aligned}
$$

As we will see next, even a small increase in wind speed results in a significant increase in power that can be extracted from the wind.

3.2.2 Available Power in the Wind

A wind turbine converts the kinetic energy of moving air to electrical energy. Of primary importance to engineers is the maximum amount of energy that can be extracted from the wind for a given wind velocity. Consider a vertical cross section of atmosphere with area A through which a steady horizontal air current of velocity, v, flows, as illustrated in Figure 3.4. To more readily identify the area with the area swept out by the blades of a horizontal axis wind turbine, the shape is shown as circular, but for purposes of this derivation, the shape is arbitrary. From basic physics, the kinetic energy of a mass, m, moving with velocity, v, is given by the relation

$$
KE = \frac{1}{2} mv^2 \tag{3.2}
$$

where the unit for KE is the joule (J). Dividing both sides of Equation (3.2) by mass, m, we obtain

$$
ke = \frac{1}{2} v^2 \tag{3.3}
$$

Figure 3.4
Diagram for calculating available wind power.

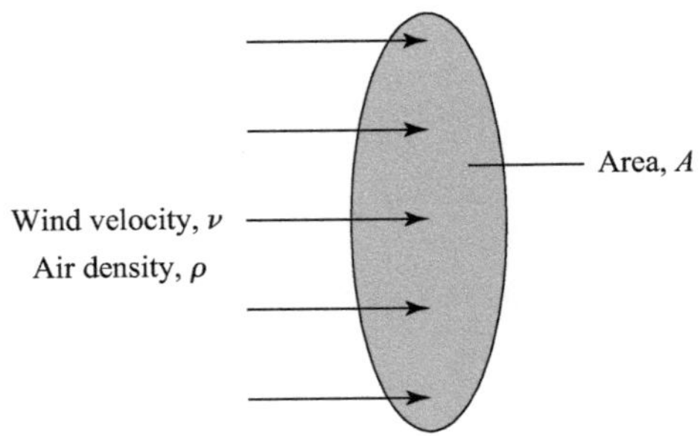

where ke is kinetic energy per unit mass, expressed in units of J/kg. The mass flow rate, $\dot{m}$, of air flowing through the area, A, is expressed as

$$\dot{m} = \rho A v \qquad (3.4)$$

where $\dot{m}$ is expressed in units of kg/s, and ρ is air density, expressed in units of kg/m^3. We see that by multiplying both sides of Equation (3.3) by mass flow rate, the left side of the equation now has units of J/s, which is defined as the watt (W), a unit of power. Hence, we have derived a relation for the available power in the wind,

$$P_{\text{wind}} = \frac{1}{2}\rho A v^3 \qquad (3.5)$$

Note that for purposes of basic wind turbine analysis presented in this chapter, wind direction is not considered. Hence, wind *speed* and wind *velocity* are used synonymously here even though speed is a scalar quantity and velocity is a vector quantity.

From Equation (3.5) we glean three important facts. First, available wind power is proportional to air density. For standard conditions at sea level, $\rho = 1.225$ kg/m^3. Second, available wind power is proportional to the area swept out by the turbine blades. This applies to any type of wind turbine. Third, and most striking, available wind power is proportional to the cube of wind speed. Hence, if wind speed doubles, the available wind power increases by a factor of eight.

Air density is a function of temperature and pressure, and both of these air properties change with elevation. Using the U.S. Standard Atmosphere model, a good approximation for air density in units of kg/m^3 is given by the relation

$$\rho = 1.225 - 1.194 \times 10^{-4} z \qquad (3.6)$$

where z, expressed in the unit of m, is elevation, or height above sea level. For Equation (3.6) to be dimensionally consistent, the number 1.225 has units of kg/m^3, and the number 1.194×10^{-4} has units of kg/m^4.

EXAMPLE 3.2

Consider a horizontal axis wind turbine in the "mile high" city, Denver, Colorado. If the diameter of the turbine blades is 20 m and the wind speed is 5.0 m/s, find the available power in the wind for this turbine.

SOLUTION

The circular area swept out by turbine blades with a 20 m diameter is

$$A = \pi D^2 / 4$$
$$= \pi (20 \text{ m})^2 / 4$$
$$= 314 \text{ m}^2$$

Denver has an elevation of approximately one mile, or 1609 m. Thus, using Equation (3.6), the air density is

$$\rho = 1.225 - 1.194 \times 10^{-4} z$$
$$= 1.225 \text{ kg/m}^3 - (1.194 \times 10^{-4} \text{ kg/m}^4)(1609 \text{ m})$$
$$= 1.033 \text{ kg/m}^2$$

Using Equation (3.5), the available power in the wind is

$$P_{\text{wind}} = \frac{1}{2} \rho A v^3$$
$$= \frac{1}{2} (1.033 \text{ kg/m}^3)(314 \text{ m}^2)(5.0 \text{ m/s})^3$$
$$= 2.03 \times 10^4 \text{ kg·m}^2/\text{s}^3 = 2.03 \times 10^4 \text{ W} = 20.3 \text{ kW}$$

Before leaving this example, let us confirm that the units on our answer are correct by breaking the watt into its base units. One watt (W) is defined as one joule per second (J/s), and one joule is defined as one newton meter (N·m). From Newton's second law, $F = ma$, we see that the newton has units of kg·m/s^2. Hence, the watt has base units of kg·m^2/s^3, as indicated in our answer.

3.2.3 The Betz Limit

The available power in the wind given by Equation (3.5) is not the maximum power that a wind turbine can extract from the wind. A wind turbine that extracts this amount of power implies that all the kinetic energy of the wind is converted to mechanical energy of a rotating shaft, yielding a wind velocity of zero downstream of the turbine blades. Obviously, the wind velocity downstream of a turbine is not zero.

Consider the schematic of a wind turbine and the streamlines of air around it, as illustrated in Figure 3.5. For purposes of this analysis, we assume that the air flow is steady (does not change with time) and incompressible (air density is constant). We also assume an infinite number of turbine blades, which is equivalent to saying that the turbine blades compose a solid disk, referred to as an actuator disk. To satisfy the conservation of mass principle, the mass flow rates of air upstream of the disk, at the disk, and downstream of the disk are equal. Thus,

$$\dot{m} = \rho A_1 v_1 = \rho A v = \rho A_2 v_2 \tag{3.7}$$

From Figure 3.5, we see that $A_1 < A_2$ and therefore $v_1 > v_2$. It can be shown that the air velocity at the actuator disk is the arithmetic average of the air velocities upstream and downstream of the actuator disk,

$$v = \frac{v_1 + v_2}{2} \tag{3.8}$$

Figure 3.5
Turbine model for the derivation of the Betz limit.

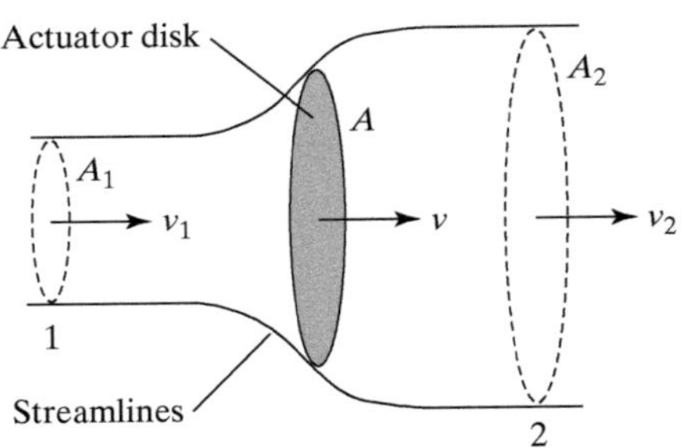

The power extracted from the wind by the turbine is the difference in the kinetic energy rates of the wind upstream and downstream of the actuator disk, which is written as

$$P = \frac{1}{2}\dot{m}(v_1{}^2 - v_2{}^2) \tag{3.9}$$

Substituting $\dot{m} = \rho A v$ into Equation (3.9), we have

$$P = \frac{1}{2}\rho A v(v_1{}^2 - v_2{}^2) \tag{3.10}$$

Substituting Equation (3.8) into Equation (3.10), we obtain

$$P = \frac{1}{2}\rho A\left(\frac{v_1 + v_2}{2}\right)(v_1{}^2 - v_2{}^2) \tag{3.11}$$

which, after algebraic manipulation, can be expressed as

$$P = \frac{1}{4}\rho A v_1{}^2\left[1 + \left(\frac{v_2}{v_1}\right) - \left(\frac{v_2}{v_1}\right)^2 - \left(\frac{v_2}{v_1}\right)^3\right] \tag{3.12}$$

Letting $x = v_2/v_1$, Equation (3.12) becomes

$$P = \frac{1}{4}\rho A v_1{}^2(1 + x - x^2 - x^3) \tag{3.13}$$

Now, we perform a maximization on the power, P, by differentiating Equation (3.13) with respect to x and setting the result to zero, which leads to the quadratic equation

$$3x^2 + 2x - 1 = 0 \tag{3.14}$$

The roots of Equation (3.14) are $1/3$ and -1. The second root is nonphysical, so we see that $v_2/v_1 = 1/3$, which means that $v_1 = 3v_2$, and therefore $v = 2v_1/3$. Substituting $x = 1/3$ into Equation (3.13) and simplifying, we obtain

$$P = \frac{16}{27}\frac{1}{2}\rho A v_1{}^3 \tag{3.15}$$

Equation (3.15) resembles Equation (3.5), the relation for the available power in the wind, except for the quantity $16/27$ on the right hand side and the subscript 1 on the quantity v to denote the velocity *upstream* of the turbine blades and not *at* the turbine blades. We conclude that the maximum theoretical power that a turbine can extract from the wind is not the available power in the wind but only

16/27 of it. Thus, we can write an expression for the maximum theoretical power that a turbine can extract from the wind as

$$P_{\text{wind, max}} = C_p P_{\text{wind}} \tag{3.16}$$

where P_{wind} is defined by Equation (3.5). The quantity C_p is called the **power coefficient**, and, as we just proved, has a maximum value of 16/27, approximately 0.593. This value is known as the **Betz limit**, in honor of the German physicist, Albert Betz, who published this result in 1919. The Betz limit is the maximum theoretical fraction of power that can be extracted from the wind, and no real turbine operates at this limit. Large wind turbines have power coefficients in the range of about 0.4 to 0.5, whereas small wind turbines have power coefficients in the range of about 0.2 to 0.4.

PRACTICE!

1. A horizontal axis wind turbine operating at sea level has a blade diameter of 40 m. Find the maximum amount of power that the turbine can extract from the wind for wind speeds of 2 m/s and 8 m/s.

   Answer: 3.65 kW, 234 kW

2. A small vertical axis wind turbine extracts 4.2 kW of power from a 7 m/s wind at sea level. If the area swept out by the blades is 60 m^2, find the power coefficient for this wind turbine.

   Answer: 0.33

3.3 TYPES OF WIND TURBINES

A *wind turbine* is a machine that converts the kinetic energy of the wind to electrical energy. As illustrated in Figure 3.6, there are two primary types of wind turbines, each being characterized by the orientation of the axis (shaft) of the turbine. A typical **horizontal axis wind turbine** (HAWT) consists of three blades mounted to a horizontal shaft. The power-generating capacity of HAWTs varies widely. As illustrated in Figure 3.7, large commercial turbines generate as much as 5 MW of power, whereas turbines used to generate power for a single home or small business generate 50 kW to 300 kW. Blade diameter varies widely also, ranging from a few meters to over 100 m. The nacelle of a HAWT is an enclosure that houses the gear box, drivetrain, electrical generator, brake system, and other mechanical and electrical components. The role of the tower is to support the turbine at a height well above the ground to intercept higher velocity winds and therefore harness more wind energy.

A **vertical axis wind turbine** (VAWT), which usually resembles an "eggbeater," typically has three blades mounted to a vertical shaft. Historically, the first wind mills were based on the vertical-axis configuration. VAWTs are less common than HAWTs and are primarily used in small-scale applications. This type of wind turbine is normally mounted on a short mast close to the ground or the roof of a building.

Wind turbines are broadly classified as either *lift* machines or *drag* machines. In lift machines, aerodynamic lift forces cause the blades to rotate, whereas in drag machines, aerodynamic drag forces cause the blades to rotate. Modern HAWTs are lift machines because the blades are designed to produce lift much like the airfoils of an aircraft. VAWTs can be either drag machines, because the blades are designed to "catch" the wind much like the cups of a weather station anemometer, or they

Figure 3.6
The two main types of wind turbines are the horizontal axis wind turbine (HAWT) and the vertical axis wind turbine (VAWT).

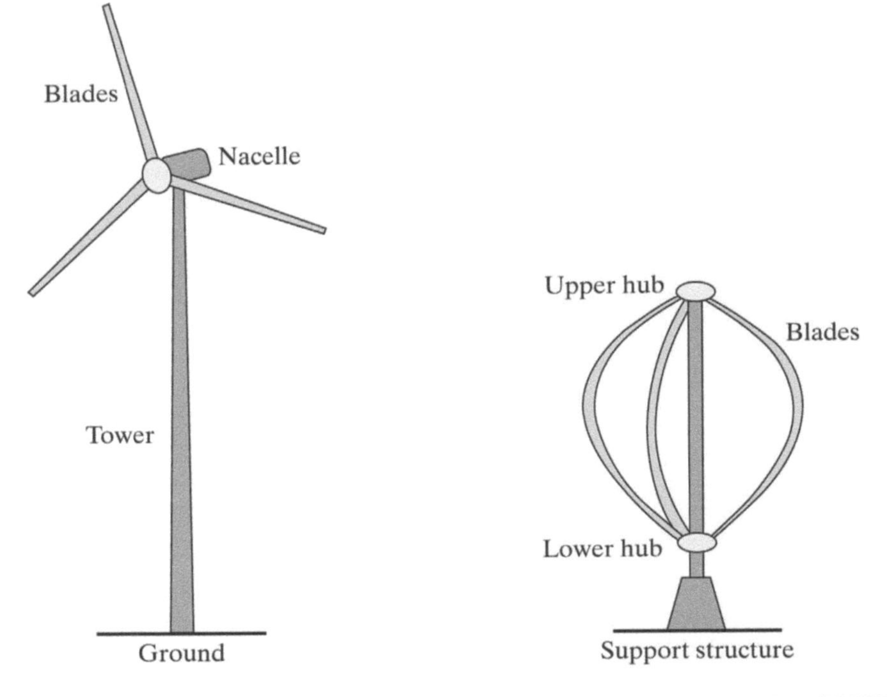

Figure 3.7
Relative power capacity, tower height, and rotor diameter of horizontal axis wind turbines.

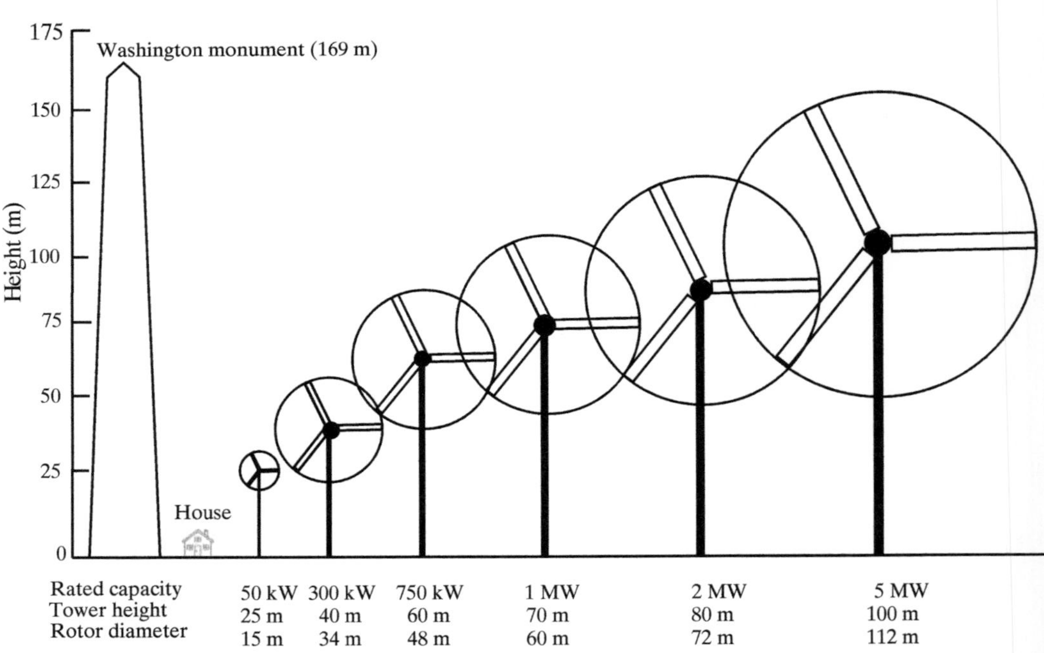

Rated capacity	50 kW	300 kW	750 kW	1 MW	2 MW	5 MW
Tower height	25 m	40 m	60 m	70 m	80 m	100 m
Rotor diameter	15 m	34 m	48 m	60 m	72 m	112 m

can be lift machines, because aerodynamic lift forces cause the blades to rotate. Basic aerodynamic analyses of lift and drag machines show that the lift forces that can be achieved by a lift machine are much greater than those that can be achieved by a drag machine with the same blade surface area.

The most fundamental decision in the selection or design of a wind turbine is the orientation of its axis, that is, whether to use a HAWT or a VAWT. HAWTs are more prevalent than VAWTs, but both have advantages and disadvantages. There are two primary advantages of HAWT systems. First, the *rotor solidity* (total blade mass relative to swept area) is lower, which minimizes the cost per kW of power generated. Second, the rotor is typically higher above the ground where wind speeds are higher. Furthermore, the efficiencies (discussed in the next section) of HAWT systems are higher, and the blades of a HAWT have the capability to self-start the turbine. The primary advantage of a VAWT is that there is no need for a yaw system since the blades will rotate under any wind direction. A second advantage of a VAWT is ease of maintenance since the gear box and generator are generally located close to the ground or roof top. Another advantage of VAWT systems is simplicity of blade design—the blades have a constant chord (width) and do not require a twist from hub to tip as do the blades of a HAWT. In spite of some advantages of VAWTs, this type of wind turbine has not gained widespread acceptance due to a variety of structural and other problems. And, as pointed out earlier, lift machines generate higher aerodynamic forces than drag machines with the same blade surface area. Hence, HAWTs are the predominant type of wind turbine in use and under development. In the next section, we discuss the HAWT further.

3.3.1 Horizontal Axis Wind Turbines

A HAWT, particularly a large commercial unit, is a complex mechanical and electrical system. For our purposes, we present a relatively brief description of the HAWT and its fundamental engineering principles. For detailed information on HAWTs, the suggested readings at the end of this chapter should be consulted.

Figure 3.8
A wind farm.

(Stephen Bures/ Shutterstock)

At the end of 2012, it is estimated that there were about 225,000 wind turbines in operation around the world, and most of these wind turbines are the horizontal axis type. Some HAWT systems operate as a single unit, supplying electrical power to a home or small business. For large-scale power-generation applications, however, HAWTs are found in a group called a **wind farm** or *wind park*. A wind farm consists of a cluster of wind turbines covering an extended area, as shown in Figure 3.8. Wind farms require open spaces with unobstructed access to the wind, making wind farms ideal for agricultural areas, grazing lands, and coastlines. A small wind farm may consist of fewer than 20 turbines covering a few acres, whereas a large wind farm may consist of hundreds of turbines covering hundreds of square miles. Most wind farms are located on land, but wind farms can also be located offshore.

The rotor of a HAWT can be either upwind or downwind of the tower, as illustrated in Figure 3.9. In order to keep the rotor pointed into the wind, an *upwind* machine requires an active yaw system consisting of a controlled motor that aligns the turbine as the wind direction changes. A *downwind* machine utilizes aerodynamic forces on the rotor, which create a moment around the tower axis, to align the turbine into the wind. A disadvantage of the downwind configuration is that the tower produces a wake behind it, so the blades must pass through the wake with every rotation. This produces periodic loads on the blades that can result in fatigue damage to them and may induce an unwanted ripple on the electrical output power. Blade passage through the wake also produces aerodynamic noise.

In a well-designed wind farm, the spacing of wind turbines is carefully considered. If the turbines are too close to one another, the power output of downwind turbines will be diminished due to interference of upwind turbines. If the turbines are spaced too far apart, the wind farm site space is underutilized. The optimum spacing between towers in the direction of the wind is between five and nine rotor diameters, and the optimum side-to-side spacing between towers is between three and five rotor diameters. This optimum spacing configuration is illustrated in Figure 3.10.

The majority of HAWT rotors have three blades. Compared to a two-blade rotor, a three-blade rotor has a slightly larger (about 5 percent) power coefficient. Three-blade machines also have a constant polar moment of inertia with respect to the yaw axis, which means that the turbine's resistance to rotation is constant about the yaw axis, giving the turbine a smoother operation. Wind turbines with more

Figure 3.9
Upwind and downwind HAWT configurations.

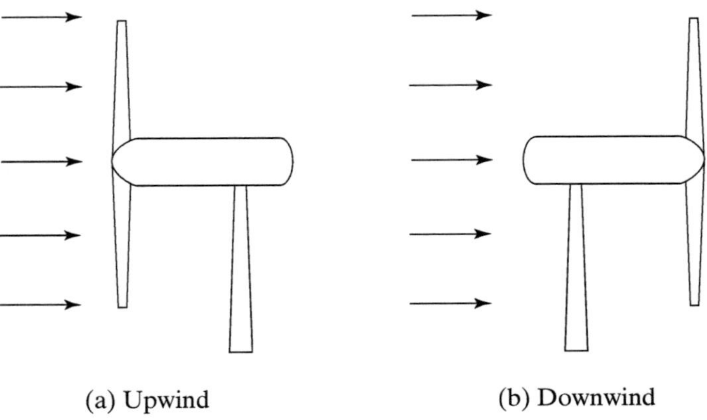

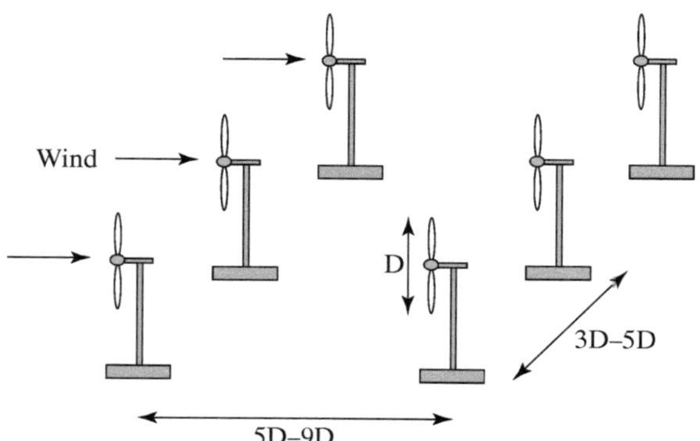

Figure 3.10
Optimum tower spacing in a wind farm.

than three blades are not optimum with respect to electrical power production and are cost prohibitive.

A simplified version of the main components of a HAWT is shown in Figure 3.11. The blades are mounted to a hub that is connected to the main shaft of the gear box. The function of the gear box is to increase the spin rate of the shaft connected to the electrical generator so that the power produced by the generator is maximized. A typical spin rate ratio is 100:1. Depending on the size and complexity of the HAWT, various types of braking systems are employed to keep the turbine speed within its allowed limits in high winds and to hold the turbine at rest during maintenance. For large turbines that have blades with controllable pitch (angle of attack), aerodynamic braking is typically used. Braking of small wind turbines can be done by dumping energy from the generator into a resistor bank, converting the kinetic energy of the shaft into heat. The function of a mechanical brake is to hold the turbine at rest while maintenance is performed. The anemometer and wind vane measure wind speed and direction, respectively, sending this data to the controller. The controller directs the yaw system, keeping the turbine pointed into the wind, and controls blade pitch and the braking system.

From the illustration of the internal components of a HAWT in Figure 3.11, it is apparent that there is sequence of power conversions that occur. Wind power is

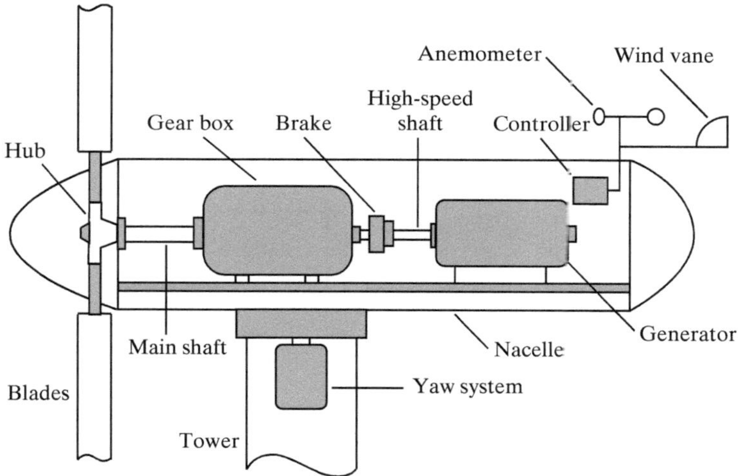

Figure 3.11
Main internal components of a HAWT.

converted to mechanical power of the main shaft. The power of the main shaft is then converted to power of the high-speed shaft, which is converted to electrical power in the generator. The combination of the turbine, gear box, and generator in which this power conversion takes place is a three-stage system called a *power train*. Due to friction and other energy loss mechanisms, some power is lost in each stage. For any power conversion system, **efficiency**, η, is the ratio of output power to input power, expressed as

$$\eta = \frac{P_{\text{out}}}{P_{\text{in}}} \qquad (3.17)$$

From the general definition of efficiency in Equation (3.17), we can express the efficiency of each stage in the power train, shown in Figure 3.12. Note that the power coefficient, C_{p}, introduced earlier is the efficiency of the turbine, defined as the ratio of main shaft power to wind power. As shown in Section 3.2.3, the maximum theoretical value of the power coefficient is 16/27, the Betz limit. The combined efficiency, η, of the wind machine is the product of the efficiencies of the turbine, gear box, and electrical generator, expressed as

$$\eta = C_{\text{p}}\eta_{\text{gb}}\eta_{\text{gen}} = \frac{P_{\text{sh}}}{P_{\text{wind}}} \frac{P_{\text{gb}}}{P_{\text{sh}}} \frac{P_{\text{elect}}}{P_{\text{gb}}} \qquad (3.18)$$

Noting the cancellation of the terms P_{sh} and P_{gb} in Equation (3.18), the combined efficiency becomes

$$\eta = \frac{P_{\text{elect}}}{P_{\text{wind}}} \qquad (3.19)$$

Hence, using Equation (3.5), the electrical output power of the wind machine may be written as

$$P_{\text{elect}} = \eta P_{\text{wind}} = \frac{1}{2} \eta \rho A v^3 \qquad (3.20)$$

Depending on the size and application, modern wind machines can generate DC (direct current) power or AC (alternating current) power. Some wind turbines generate DC power, which is inverted to AC power to match the frequency of the power grid.

Figure 3.12
Power train and typical efficiencies of a HAWT.

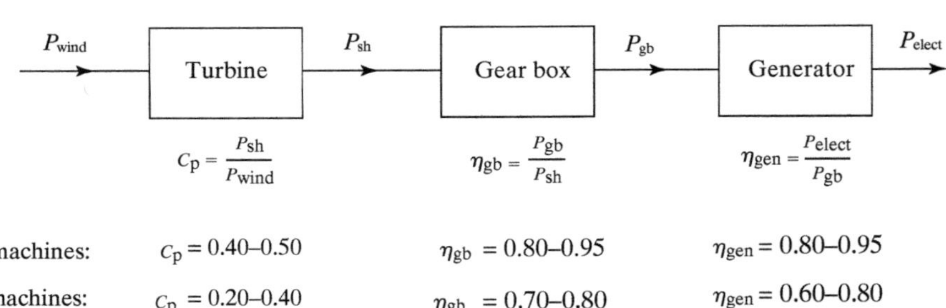

EXAMPLE 3.3

A HAWT in Laramie, Wyoming, has a blade diameter of 30 m. The power coefficient, gear box efficiency, and generator efficiency are 0.45, 0.85, and 0.90, respectively. For a wind speed of 6 m/s, find the electrical output power.

SOLUTION

Laramie, Wyoming, has an elevation of 2184 m. Using Equation (3.6), air density is

$$\rho = 1.225 - 1.194 \times 10^{-4}z$$
$$= 1.225 \text{ kg/m}^3 - (1.194 \times 10^{-4} \text{ kg/m}^4)(2184 \text{ m})$$
$$= 0.9642 \text{ kg/m}^3$$

The circular area swept out by turbine blades with a 30 m diameter is

$$A = \pi D^2/4$$
$$= \pi(30 \text{ m})^2/4$$
$$= 707 \text{ m}^2$$

Using Equations (3.18) and (3.20), the electrical output power is

$$P_{\text{elect}} = \frac{1}{2}\,\eta\rho A v^3$$

$$= \frac{1}{2}(0.45)(0.85)(0.90)(0.9642 \text{ kg/m}^3)(707 \text{ m}^2)(6 \text{ m/s})^3$$

$$= 2.53 \times 10^4 \text{W} = 25.3 \text{ kW}$$

EXAMPLE 3.4

A site for a wind farm near Redding, California, has a plot of land that can accommodate 300 HAWTs. The required power-generation capacity of the wind farm is estimated to be 25 MW. Using $C_p = 0.50$, $\eta_{\text{gb}} = 0.80$ and $\eta_{\text{gen}} = 0.90$, and assuming that all wind turbines in the farm are identical, find the required length for the turbine blades. Use a design wind speed of 7 m/s.

SOLUTION

Redding, California, has an elevation of 151 m. Using Equation (3.6), air density is

$$\rho = 1.225 - 1.194 \times 10^{-4}z$$
$$= 1.225 \text{ kg/m}^3 - (1.194 \times 10^{-4} \text{ kg/m}^4)(151 \text{ m})$$
$$= 1.207 \text{ kg/m}^3$$

Expressing Equation (3.20) in terms of diameter, D, of the circle swept out by the blades and the number of turbines, N, in the wind farm, we have

$$P_{\text{elect, farm}} = \frac{1}{2}\,N\eta\rho(\pi D^2/4)\,v^3$$

(continued)

Rearranging and solving for D, we obtain

$$D = \left(\frac{8P_{\text{elect, farm}}}{N\eta\rho\pi v^3}\right)^{1/2}$$

$$= \left[\frac{8(25 \times 10^6 \, \text{W})}{(300)(0.36)(1.207 \, \text{kg/m}^3)\pi(7 \, \text{m/s})^3}\right]^{1/2}$$

$$= 37.7 \, \text{m}$$

Neglecting the diameter of the rotor hub, the blade length, L, is half the diameter, D, of the circle swept out by the blades. Thus, we have

$$L = D/2$$

$$= (37.7 \, \text{m})/2$$

$$= 18.9 \, \text{m}$$

It should be noted that 900 blades (3 blades/turbine $\times$ 300 turbines) are required for this wind farm.

EXAMPLE 3.5

A small HAWT for augmenting grid power to a seaside residence has a blade diameter of 4.5 m. The required power capacity of the turbine is 5 kW. Using the efficiencies at the upper limit of the ranges given in Figure 3.12, find the wind speed corresponding to this power capacity.

SOLUTION

The elevation of a seaside home is 0 m, so air density is $\rho = 1.225 \, \text{kg/m}^3$. The circular area swept out by the turbine blades is

$$A = \pi D^2/4$$

$$= \pi(4.5 \, \text{m})^2/4$$

$$= 15.9 \, \text{m}^2$$

Using the values of efficiencies at the upper limits of small wind turbines, the combined efficiency is

$$\eta = C_{\text{p}}\eta_{\text{gb}}\eta_{\text{gen}}$$

$$= (0.40)(0.80)(0.80)$$

$$= 0.256$$

Using Equation (3.20) and solving for v, we obtain

$$v = \left(\frac{2P_{\text{elect}}}{\eta\rho A}\right)^{1/3}$$

$$= \left[\frac{2(5000 \, \text{W})}{(0.256)(1.225 \, \text{kg/m}^3)(15.9 \, \text{m}^2)}\right]^{1/3}$$

$$= 12.6 \, \text{m/s} \, (28.2 \, \text{mi/h})$$

PRACTICE!

1. A HAWT in Fargo, North Dakota, experiences winds at an average speed of 8 m/s. The power coefficient, gear box efficiency, and generator efficiency are 0.40, 0.82, and 0.94, respectively. If the turbine blade length is 26 m, find the electrical output power.

   Answer: 200 kW

2. Compare the electrical output power of two identical HAWTs subjected to the same wind speed, one operating in San Diego, California, ($z = 0$ m) and the other in Santa Fe, New Mexico, ($z = 2210$ m) by calculating the output power ratio of the San Diego machine to the Santa Fe machine.

   Answer: 1.27

APPLICATION—CALCULATING THE ELECTRICAL ENERGY GENERATION OF A WIND TURBINE

In Section 3.3 we showed how to find the *instantaneous* electrical output power of a HAWT, that is, the electrical output power of a turbine at a moment of time for a given wind speed. Of particular interest to engineers who design and analyze wind turbines is the *total amount of electrical energy* that a wind turbine generates during a specific time period. This time period may be a day, a month, a year, or some other prescribed period of time. By finding the total electrical energy produced by a wind turbine (or wind farm) during a given time period, engineers can more fully assess the degree to which the wind system meets the long-term electrical energy demands of a building or community.

As discussed in Section 3.2.1, wind speeds vary with time. It would be useful to have a straightforward method for calculating the total electrical energy generated by a turbine for a period of time during which wind speed variations are taken into account. Recall that power is the time rate of energy, defined by the formula

$$P = E/\Delta t \tag{3.21}$$

where P is power, E is energy, and Δt is a time interval Hence, the electrical energy, E_{elect}, generated by a wind turbine during a time interval Δt is

$$E_{elect} = P_{elect}\Delta t \tag{3.22}$$

where P_{elect}, given by Equation (3.20), is the electrical output power corresponding to the time interval, Δt, during which the wind speed is considered to be constant, or at least an average value taken over the time interval. To apply Equation (3.22) to an extended time period, such as a day, week or month, we sum the product of electrical output power and time interval over the desired number of time intervals, N, yielding the relation

$$E_{elect} = \sum_{i=1}^{N} P_{elect,i}\Delta t_i \tag{3.23}$$

Typically, the time interval, Δt, is chosen to be constant across the extended time period considered, so Equation (3.23) may be written as

$$E_{\text{elect}} = \Delta t \sum_{i=1}^{N} P_{\text{elect},i} \tag{3.24}$$

To illustrate the method, consider a HAWT operating at sea level during a 24-hour period. The diameter of the turbine blades is 35 m and the combined efficiency of the turbine is 0.28. Average wind speeds over the 24-hour period are shown in Table 3.2. In practice, these average wind speeds would be based on measurements taken over a suitably long period of time. At this particular wind site, wind speeds gradually increase during the early morning hours and then diminish in the afternoon. Note that for wind speeds under 3 m/s, the power generation is zero. Wind turbine blades require a certain amount of torque before they begin to rotate. The wind speed at which this torque is reached is called *cut-in* speed. A typical cut-in speed range is 6 to 8 mi/h

Table 3.2 Summary of Wind Turbine Electrical Energy Calculations

Time interval	Average wind speed (m/s)	P_{elect} (kW)	E_{elect} (kWh)
12 am–1 am	0.6	0.0	0.0
1 am–2 am	0.8	0.0	0.0
2 am–3 am	1.9	0.0	0.0
3 am–4 am	2.9	0.0	0.0
4 am–5 am	3.5	7.1	7.1
5 am–6 am	5.2	23.2	23.2
6 am–7 am	7.6	72.4	72.4
7 am–8 am	9.3	132.7	132.7
8 am–9 am	12.4	314.6	314.6
9 am–10 am	15.8	650.8	650.8
10 am–11 am	16.2	701.5	701.5
11 am–12 noon	15.4	602.6	602.6
12 noon–1 pm	13.1	370.9	370.9
1 pm–2 pm	11.0	219.6	219.6
2 pm–3 pm	10.6	196.5	196.5
3 pm–4 pm	9.6	146.0	146.0
4 pm–5 pm	8.4	97.8	97.8
5 pm–6 pm	6.7	49.6	49.6
6 pm–7 pm	5.4	26.0	26.0
7 pm–8 pm	3.1	4.9	4.9
8 pm–9 pm	0.7	0.0	0.0
9 pm–10 pm	0.6	0.0	0.0
10 pm–11 pm	0.5	0.0	0.0
11 pm–12 am	0.5	0.0	0.0
			Total = 3616 kWh

(3 m/s to 4 m/s). Using Equations (3.20) and (3.24), we calculate the total electrical energy generated by the turbine for the 24-hour period. Note that the values of power, in units of kW, and energy, in units of kWh, for a given time interval are numerically equivalent because the time interval is precisely one hour. By adding the energy values in the last column, we obtain 3616 kWh. Because of the repetitive nature of these calculations, the use of a spreadsheet is suggested.

There are more sophisticated methods, such as the *method of bins*, for calculating the energy generation of wind turbines over extended time periods. These methods employ statistics for modeling wind speed variations.

3.3.2 Vertical Axis Wind Turbines

As pointed out in Section 3.3, HAWTs are more predominant than VAWTs. However, VAWTs are interesting wind machines and deserve some additional discussion.

VAWTs are divided into two categories, drag machines and lift machines. In a drag machine, the blades are designed to produce high aerodynamic drag forces in order to rotate the vertical shaft. Some basic blade designs of drag machines are illustrated in Figure 3.13. In a paddlewheel machine, flat paddle wheels are mounted to a series of spokes symmetrically connected to the vertical shaft. A screen shields half of the machine from the wind so that the drag forces produce a net moment on the shaft, causing it to rotate. The Savonius machine, named after the Finnish engineer, Sigurd Johannes Savonius, consists of one or more S-shaped blades that resemble a scoop. Drag forces are greater for air flowing into the concave sides of the blades than the convex sides, thereby producing a net moment on the shaft. Cupped machines resemble the familiar anemometers used by meteorologists for measuring wind speeds.

In a lift machine, the blades are designed to produce aerodynamic lift similar to that of the wings of an aircraft. Some basic blade designs of lift machines are illustrated in Figure 3.14. The lift type of VAWT that is the most competitive with the HAWT is the Darrieus machine, invented by the French engineer, Georges Jean Marie Darrieus. The Darrieus machine typically consists of two or three curved blades attached to the top and bottom of the vertical shaft, giving this type of VAWT an "eggbeater" shape. Two other blade designs, the horizontal-vertical axis wind turbine (H-VAWT) and the vertical-vertical axis wind turbine (V-VAWT), have

Figure 3.13
Blade designs of drag machines (top view).

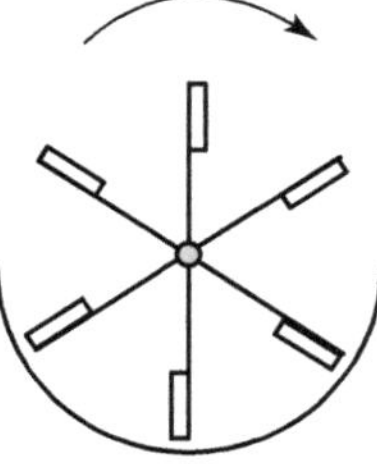
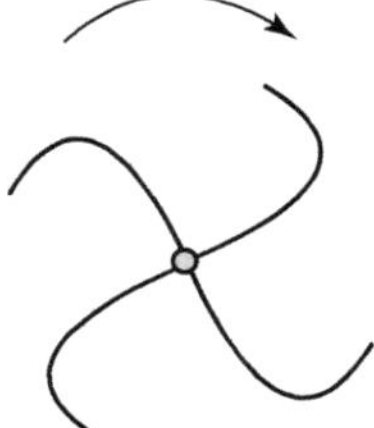
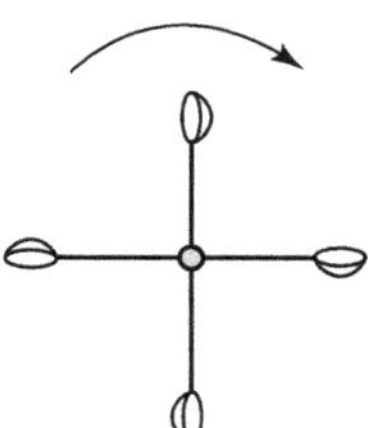

Screened paddlewheel   Savonius (multibladed)   Cupped (anemometer)

Figure 3.14
Blade designs of lift
machines (side view).

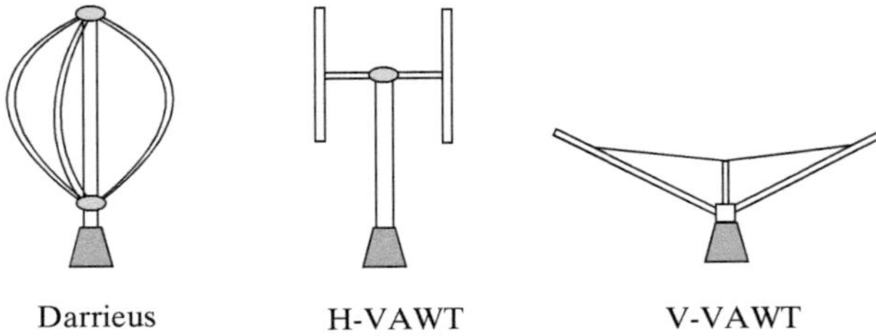

been investigated. As seen in Figure 3.14, the H-VAWT blades form an "H" and the V-VAWT blades form a "V." A Darrieus VAWT is shown in Figure 3.15.

The analysis of a VAWT is fundamentally the same as the analysis of a HAWT. The electrical output power of a VAWT is given by Equation (3.20). However, the area, A, in Equation (3.20) is not the area of a circle, as can be ascertained in Figures 3.13 and 3.14. For a vertical axis turbine, the area, A, is the "presented" or "frontal" area of the blades. In other words, A is the area "seen" by the air as it approaches the blades. For a paddlewheel and Savonius machine, the area is the product of rotor height and diameter. Similarly, for the H-VAWT, the area is the product of rotor height and diameter, but for the V-VAWT, the area is the "V"-shaped area swept out by the blades. For the Darrieus machine, the area depends on the shape of the arc formed by the blades. This arc may be approximated by a conic section (ellipse, parabola, or hyperbola) for the purpose of calculating the area, A. If an ellipse is used, the presented area of the blades is given by the approximation $A \approx 0.65DH$, where D and H are the diameter and height of the rotor, respectively. The following example shows how to find the area for parabolic-shaped blades.

Figure 3.15
A Darrieus vertical axis
wind turbine.

(Spirit of America/
Shutterstock)

EXAMPLE 3.6

A Darrieus wind machine operating at sea level has two blades in the shape of a parabola, as illustrated in Figure 3.16. The wind speed is 10 m/s, and the combined efficiency of the machine is 0.24. If the parabolic function is $y = x^2/4$, find the electrical output power.

SOLUTION

First, we calculate the area, A, swept out by the turbine blades. The blades have a parabolic shape that follows the function $y = x^2/4$. We construct a graph of the parabola, conveniently oriented with respect to the x and y axes. Using a differential area of $dA = (4 - y)\,dx = (4 - x^2/4)\,dx$, we integrate to find the area enclosed by one blade and the shaft, indicated by the dashed line in Figure 3.16. Noting that the area swept out by the blades is twice this area, we write

$$A = 2\int_{-4}^{4} dA = 2\int_{-4}^{4} (4 - x^2/4)\,dx = 2(4x - x^3/12)\big|_{-4}^{4}$$
$$= 42.7 \text{ m}^2$$

Using Equation (3.20), the electrical output power is

$$P_{\text{elect}} = \frac{1}{2}\eta\rho A v^3$$

$$= \frac{1}{2}(0.24)(1.255 \text{ kg/m}^3)(42.7 \text{ m}^2)(10 \text{ m/s})^3$$

$$= 6.43 \times 10^3 \text{W} = 6.43 \text{ kW}$$

Note that if we assume an elliptical shape for the blades, keeping the same rotor height and diameter, the area swept out by the blades is

$$A \approx 0.65DH = 0.65(8 \text{ m})(8 \text{ m}) = 41.6 \text{ m}^2$$

which is very close to the area for parabolic-shaped blades.

Figure 3.16
Darrieus wind machine and blade shape for Example 3.6.

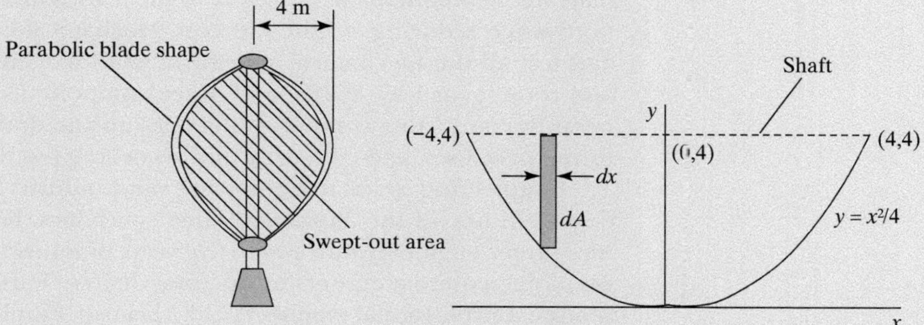

PRACTICE!

1. A H-VAWT operating in Boise, Idaho, has a rotor height and diameter of 7.5 m and 6.0 m, respectively. The combined efficiency of the machine is 0.27. For a wind speed of 9 m/s, what is the electrical output power?

   Answer: 4.99 kW

2. An experimental wind machine in Seattle, Washington, has a combined efficiency of 0.18. The machine is a V-VAWT with a rotor height and diameter of 3.0 m and 7.0 m, respectively. Find the electrical output power for a wind speed of 12 m/s.

   Answer: 2.00 kW

PROFESSIONAL SUCCESS—WIND ENERGY: AN INTERDISCIPLINARY INDUSTRY

Engineers who work in the wind energy industry come from a variety of educational backgrounds and disciplines. Wind machines are complex systems that require a broad range of engineering knowledge and skills to develop and build them. Aerospace and aeronautical engineers design, analyze, and test turbine blades and rotors and work with meteorologists in the site selection of wind turbines. Civil engineers design and supervise the construction of wind farms and the surrounding infrastructure, including roads and support buildings. Civil engineers also design wind turbine towers and foundations and conduct geotechnical studies on wind machine sites. Electrical engineers design, analyze, and test turbine motors, controls, generators, and transmission lines. Environmental engineers conduct assessments of the environmental impacts of wind turbines, such as effects on birds, noise generation, visual impacts, and telecommunications interference. Safety engineers identify potential hazards of wind machines and implement measures to assure that safe manufacturing and operating procedures are followed. Manufacturing and industrial engineers optimize manufacturing methods and operations to maximize productivity and minimize costs. Materials engineers develop and test materials used for turbines blades, nacelles, towers, and other wind machine components in an effort to meet structural and electrical specifications while reducing weight and cost. Mechanical engineers design, analyze, and test all the mechanical systems of wind machines, including the blades and rotor, gear box, shafts, and other components. Technicians assist engineers by conducting computer-aided design and drafting, setting up manufacturing processes, and collecting and recording test data.

Engineering graduates enter the wind industry with at least a bachelor's degree in one of the aforementioned specialties. Some wind energy companies prefer engineers with two to five years of industrial experience, and some positions require a masters or doctoral degree. Furthermore, some positions require a professional engineers (PE) license. Employment in wind energy is projected to increase significantly in the coming years to meet the demands of this growing renewable energy industry.

3.4 ENVIRONMENTAL CONSIDERATIONS

Wind machines do not release harmful emissions into the atmosphere. This does not mean, however, that wind machines do not pose any negative environmental problems. All renewable energy systems, including wind, negatively impact the environment in some way. There are five primary categories of environmental impacts associated with wind energy:

1. Bird interactions
2. Wind turbine noise
3. Visual impacts
4. Electromagnetic interference of turbine blades
5. Land use

The problem of bird interactions with wind turbines first appeared in the United States during the 1980s when it was discovered that golden eagles and red-tailed hawks were being killed by wind turbines and high-voltage transmission lines in wind farms in Altamont Pass near Livermore, California. As wind energy advances and wind farms spread across the United States and the world, bird kills continue to be a problem. Bird kills by collisions with turbine blades and electrocution by transmission lines are not the only issues. Wind turbines also have the potential to change bird foraging and migration patterns, reduce available habitat, and disturb breeding and nesting. Studies that address bird interaction issues continue in an effort to mitigate these problems.

Noise from a wind turbine, or any source for that matter, is measurable and can therefore be quantified. However, noise is also a subjective matter that depends on the perceptions and opinions of people. Furthermore, ambient noises from wind rushing through trees, birds, vehicular traffic, and commercial and industrial operations determine the degree to which wind turbine noise is noticeable or annoying. Thus, what is an acceptable noise level in an urban region may be unwanted in a rural area.

Wind turbines generate two kinds of noise—aerodynamic noise from rotating turbine blades and mechanical noise from rotating machinery. Aerodynamic noise, the more severe of the two noise sources, is a complex fluid mechanics phenomenon that consists of several physical mechanisms, including the passage of the blades through the tower wake, interaction of blades with atmospheric turbulence, interaction of tip turbulence with the blade tip surface, vortex shedding at the blade trailing edge, and nonlinear boundary layer instabilities. Aerodynamic noise has been reduced in modern turbine blades by optimizing angle of attack, modifying trailing edges, and other schemes. Mechanical noise is generated by the gear box, generator, yaw drive, cooling fans, and hydraulic systems. Depending on the location of the mechanical noise source, noise is either emitted directly from the source to the air or transmitted through structural components and then emitted to the air. Mechanical noise can be mitigated by adding acoustic insulation or baffles, using specially finished gear teeth and employing vibration isolators and soft mounts for large mechanical components.

Of all the environmental impacts of wind energy, visual impact is the most difficult to quantify because it depends on individual perceptions, which may be depend on familiarity with renewable technologies, views on fossil fuels or concerns about property values. The rotating blades of a wind turbine can cast moving shadows, causing a flickering effect. Similarly, rotating turbine blades can cause a similar flickering result by reflecting sunlight, creating a flashing effect. Both of these light phenomena are visually distracting and can even affect people with epilepsy. Good design of wind turbines and wind farms includes consideration of visual elements such as number and arrangement of wind turbines, turbine type, number of blades, and tower height and color.

Electromagnetic waves generated by a radio, television or microwave transmitter can be reflected, scattered or diffracted by a wind turbine. The physical parameters of wind turbines that affect electromagnetic interference include type of wind turbine, turbine dimensions, blade rotational speed, blade geometry, tower geometry, and blade material. Of these parameters, blade rotational speed and blade material are the most critical. Older HAWTs with metal blades have caused interference of television signals near the turbines. Modern wind turbine blades are constructed of composite materials that do not cause electromagnetic interference and are structurally superior to metal blades.

Wind energy impacts land use in a number of ways. Primary issues include the actual land required per energy output, amount of land disturbed by a wind farm, nonexclusive land use, preservation of rural space, turbine density in wind farms, impact of access roads, and erosion control. Wind farms are considered to be more land intensive than other types of power plants in the sense that wind farms require more land per unit of power generated than other power plants. Wind farms can occupy from 10 to 80 acre per megawatt of installed power capacity. However, even though a large parcel of land is required for a wind farm, the footprint of a wind turbine is small compared to the total land area. Wind farms are limited to regions with consistent wind sources over extended periods of time. In the United States, wind farms are primarily located in rural areas where the land is used for agricultural, recreational or scenic purposes.

SUMMARY

Wind is one of the oldest renewable energy sources used by man. The objective of wind power systems is to convert the kinetic energy of moving air to electrical energy, which is accomplished by means of a wind turbine.

Wind speeds vary with both space and time. Spatial variations of wind speed range from centimeters to thousands of kilometers, and temporal variations of wind speed range from seconds to years. These variations must be taken into account in the site selection and design of wind power systems.

The available power in the wind is proportional to air density, area swept out by the blades, and wind speed cubed. The Betz limit, which applies to all turbines, gives the maximum theoretical fraction, 16/27, of available wind power that a turbine can extract from the wind. This quantity is known as the power coefficient. Real power coefficients are less than the Betz limit.

There are two primary types of wind turbines: horizontal axis and vertical axis. HAWTs, the more prevalent of the two, vary in power capacity from about 50 kW to 5 MW. Wind turbines are classified as either lift machines or drag machines. HAWT are lift machines, but most VAWTs are drag machines. Wind turbines installed as a group constitutes a wind farm or wind park.

Efficiency is the ratio of output power to input power. The combined efficiency of a wind turbine is the product of turbine efficiency (power coefficient), gear box efficiency and electrical generator efficiency. The electrical energy generated by a wind turbine during a given time interval is found by multiplying the power by the time interval over which wind speed is assumed constant or averaged.

The five primary categories of environmental impacts are bird interactions, wind turbine noise, visual impacts, electromagnetic interference of turbine blades, and land use.

KEY TERMS

Betz limit
boundary layer
efficiency

horizontal axis wind
  turbine
power coefficient
wind farm

wind turbine
vertical axis wind
  turbine

SUGGESTED READING

BOYLE, G., Ed., *Renewable Energy–Power for a Sustainable Future* 3rd Ed., Oxford: Oxford University Press, 2012.

BREEZE, P., *Power Generation Technologies*, Burlington, MA: Elsevier, 2005.

GIPE, P., *Wind Power–Renewable Energy for Home, Farm, and Businesses*, White River Junction, VT: Chelsea Green Publishing Company, 2004.

HAGEN, K.D., *Introduction to Engineering Analysis* 4th Ed., Upper Saddle River, NJ: Prentice Hall, 2014.

KREITH, F., *Principles of Sustainable Energy Systems* 2nd Ed., Boca Raton, FL: CRC Press, 2014.

MANWELL, J.F., J.G. McGOWAN, and A.L. ROGERS, *Wind Energy Explained–Theory, Design and Application*, West Sussex, UK: John Wiley & Sons, 2003.

SHEPHERD, W., and L. ZHANG, *Electricity Generation Using Wind Power*, Hackensack, NJ: World Scientific, 2011.

ZOOBA, A.F., and R.C. BANSAL, Eds., *Handbook of Renewable Energy Technology*, Hackensack, NJ: World Scientific, 2011.

PROBLEMS

For the following problems, it is recommended that you use the general analysis procedure of (1) problem statement, (2) diagram, (3) assumptions, (4) governing equations, (5) calculations, (6) solution check, and (7) discussion. This procedure is covered in Chapter 3 of Hagen, K.D., *Introduction to Engineering Analysis* 4th Ed.

The Wind Energy Source

3.1 At a height of 4 m above the ground, the wind speed is 2.5 m/s. Estimate the wind speeds at heights of 25 m and 50 m if the ground is covered with tall grass.

3.2 In a large city with tall buildings, the wind speed at 8 m above the ground is 2 m/s. At what height is the wind speed approximately 5.5 m/s?

3.3 Find the elevations of Chicago, Denver, Las Vegas, Helena and Topeka, and calculate the air densities for each city.

3.4 For the cities in Problem 3.3, find the available power in the wind for a HAWT with a blade diameter of 40 m. What is the maximum theoretical amount of power that the turbine can extract from the wind for these cities? For all calculations, use a wind speed of 13 m/s.

3.5 Construct a plot of the maximum theoretical power that a wind turbine can extract from the wind per unit area as a function of wind speed. Use a wind speed range of 0 m/s to 30 m/s, and assume a sea-level air density of 1.225 kg/m^3.

3.6 The desired power capacity of a wind turbine for a wind speed of 20 m/s is 250 kW. If the power coefficient is 0.40, find the required area swept out by the turbine blades. Let $\rho = 1.225 \text{ kg/m}^3$.

3.7 A HAWT has a blade diameter of 18 m. For the region where the wind turbine is to be installed, wind speed never exceeds 16 m/s. If the required maximum theoretical power that the turbine can extract from the wind is 350 kW, can this turbine be installed in Santa Fe, New Mexico? San Diego, California? Explain.

Horizontal Axis Wind Turbines

3.8 A HAWT has a blade diameter of 40 m. The power coefficient, gear box efficiency, and generator efficiency are 0.42, 0.83, and 0.91, respectively. If the wind speed is 17 m/s, calculate the electrical output power if the turbine operates in Reno, Nevada.

3.9 A homeowner in Raleigh, North Carolina, wants to install a wind turbine in his back yard to augment electrical power from the grid. A local supplier states that the combined efficiency of their turbines is approximately 0.28. How much electrical power can this turbine generate for a wind speed of 18 m/s if the turbine blade diameter is 6 m?

3.10 A certain coastal region experiences relatively constant light winds 24 hours a day. How much electrical energy can a HAWT generate in one day in this region if the wind speed is 5 m/s? The blade diameter is 22 m, and the combined efficiency is 0.30.

3.11 A site for a wind farm near Spokane, Washington, has a plot of land that can accommodate 200 HAWTs. The combined efficiency of the turbine is 0.19. Using a design wind speed of 8 m/s, find the required length for the turbine blades if the required power-generation capacity of the wind farm is 30 MW. Assume that all turbines are identical.

3.12 A small HAWT in Albany, New York, has a blade diameter of 6.2 m. If the combined efficiency of the turbine is 0.21, find the wind speed required to achieve an electrical output power of 6 kW.

3.13 Based on measurements taken over a one-hour period in Lincoln, Nebraska, wind speed varies according to the data given in Table P3.13. Find the total electrical energy generated during this period by a HAWT with a blade diameter of 20 m. The combined efficiency of the turbine is 0.25.

Table P3.13 Wind Speeds for Problem 3.13

Time period (min)	Wind speed (m/s)
5	6
2	8
8	12
4	14
2	11
6	8
5	5
8	5
4	7
5	14
5	15
6	10
Total time = 60 min	

3.14 A rancher near Salem, Oregon, takes hourly wind speed measurements over a 24-hour period, yielding the graph shown in Figure P3.14. The specifications of his HAWT state that the turbine has a power coefficient of 0.43, a gear box efficiency of 0.75, a generator efficiency of 0.88, and a blade diameter of 15 m. Using the rancher's wind speed measurements, calculate the electrical energy that the wind machine will generate during a 24-hour period.

Figure P3.14

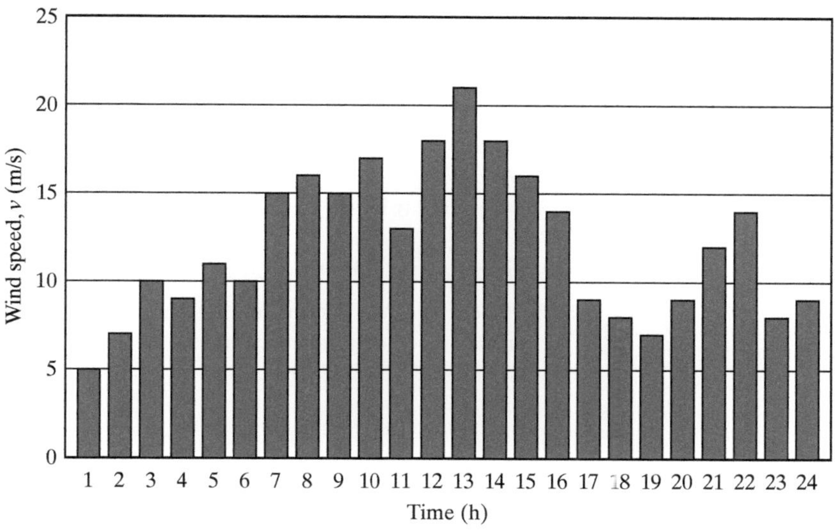

3.15 A wind farm near Green River, Wyoming, is designed to have 150 HAWTs, each with a combined efficiency of 0.34 and a blade diameter of 60 m. Based on wind speed measurements taken over a three-year period, average monthly wind speeds are obtained. Use the average monthly wind speeds to estimate the annual electrical energy generation of the wind farm. A graph of the data is given in Figure P3.15.

Figure P3.15

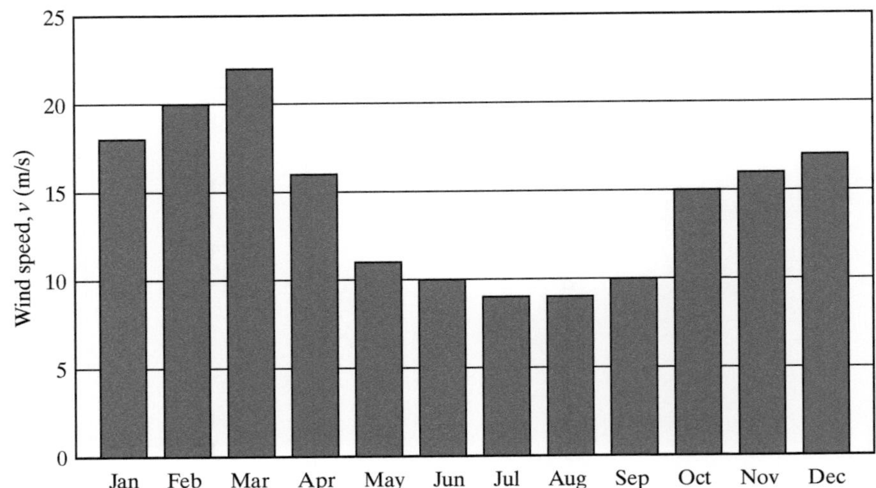

3.16 A residential community near a mountain pass with steady winds requires an electrical power capacity of 100 MW. Planners propose to install a wind farm in the mountain pass to supply one half of this power. The diameter of the turbine blades is 50 m, and the combined efficiency of the wind machines is 0.29. If the elevation of the mountain pass is 600 m, how many wind machines are required if 16 m/s is used as a design wind speed?

3.17 A homeowner wishes to have a HAWT installed on his property to offset the cost of electrical power furnished by the local electrical utility. The anticipated annual energy production of the turbine is 20,500 kWh/y, and the energy cost of electricity in his community is $0.085/kWh. If the homeowner desires a simple payback of 6 years, what is the initial cost?

3.18 A wind farm with an installed cost of $150 million has a power-generation capacity of 60 MW. The fixed charge rate is 6.5 percent, the plant capacity factor is 0.38 and the annual operation and maintenance is 1.25 percent of the initial cost. If the levelized replacement cost is averaged over a 25-year lifetime, what is the cost of energy?

Vertical Axis Wind Turbines

3.19 The H-VAWT shown in Figure P3.19 has a combined efficiency of 0.21. If the turbine operates in a coastal region, find the electrical output power for a wind speed of 18 m/s.

Figure P3.19

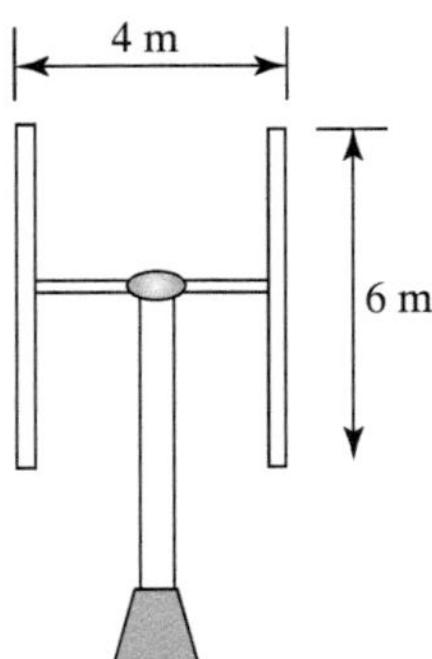

3.20 The Darrieus VAWT shown in Figure P3.20 has elliptical-shaped blades. The turbine operates at a sea-level location and has a combined efficiency of 0.24. For a wind speed of 14 m/s, find the electrical output power.

Figure P3.20

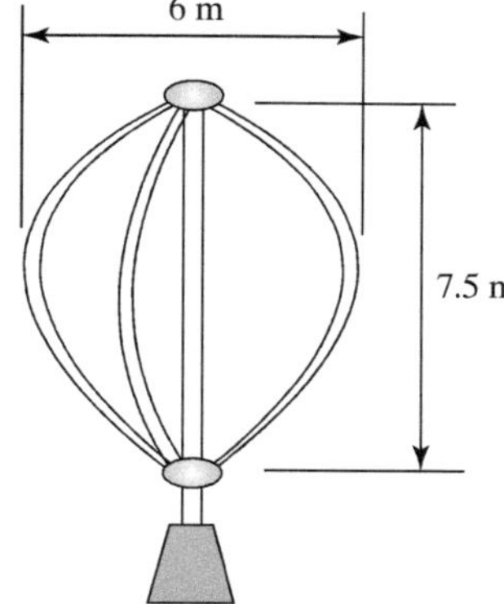

3.21 A Savonius wind turbine, shown in Figure P3.21, is located in a region where the elevation is 1300 m. For a wind speed of 16 m/s and a combined efficiency of 0.20, what blade height, H, is required for an electrical output power of 45 kW?

Figure P3.21

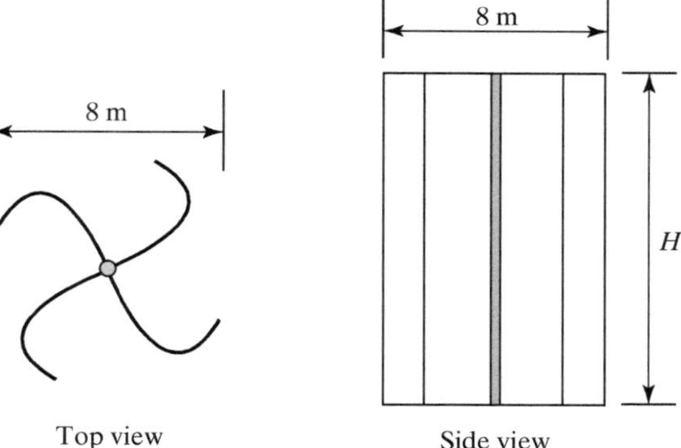

3.22 An inventor proposes to retrofit utility poles with small Savonius wind turbines, providing some electrical power for street and parking lights. The height and diameter of the blades are 2.5 m and 1.75 m, respectively. The combined efficiency of the wind machine is 0.18. Assuming a wind speed of 6 m/s, find the electrical output power of the turbine. Assume sea-level operation. Is this proposal viable?

3.23 Using the vertical axis wind turbine designs discussed in this chapter for idea generation, design your own VAWT. Make a sketch or drawing of the blade configuration, and explain how your design works.

Environmental Considerations

3.24 Research a wind farm that has been in operation for at least ten years and has experienced a pattern of bird injuries and deaths. Write an essay on your research.

3.25 Write an essay on land use of large commercial wind farms.

3.26 Write an essay that compares the environmental impacts of wind farms on land and off shore.

3.27 Write an essay on the measures that can be taken to reduce noise from wind turbines.

4 Hydroenergy

Objectives

After reading this chapter, you will have learned:

- What hydroenergy is
- The capacity of hydroenergy to supply global electricity demand
- About the basic types of hydroelectric plants
- How to find the available power in water
- How to do a basic analysis of a hydroelectric plant
- About the two basic types of turbine wheels
- The environmental impacts of hydroenergy

4.1 INTRODUCTION

Of all the renewable energy sources, hydro is the oldest. The use of waterpower dates back to ancient Egypt and Mesopotamia where water wheels powered mills for grinding grain and saws for cutting wood and stone. During the Middle Ages water wheels were used to pump water from mines and to power machines for forging metals. Other uses of waterpower were textile manufacturing, husking rice, papermaking, and pulping of sugar cane. Even though steam became the major source of energy during the industrial revolution, the use of waterpower continued well into the nineteenth century.

The invention of the electrical generator in the late nineteenth century presented a new way to harness the energy of water. By connecting water turbines to generators, a dependable source of electricity was created that could be used to power factories, businesses, and streetlights. The rivers of the eastern United States were rapidly exploited to provide electricity for an expanding industrial society. The first hydroelectric plant was constructed in 1881 at Niagara Falls, the falls that straddle the border between the state of New York and Ontario, Canada. The Niagara Falls plant originally supplied electricity for streetlights. By the end of the 1880s, over 200 additional hydroelectric plants had been constructed in the United States.

One of the most renowned hydroelectric facilities in the United States is the Hoover Dam, formerly called Boulder Dam, which impounds water in the Black Canyon of the Colorado River on the border between Nevada and Arizona. The Hoover Dam, shown in Figure 4.1, was completed in 1936 and provides electricity to utility companies in Nevada, Arizona, and California.

The terms **hydropower** and **hydroelectric** are used synonymously to describe the generation of electrical power from flowing or stored water. The kinetic energy of water flowing in a river or stream can be harnessed directly to produce work. More commonly, the flow of water is regulated by means of a reservoir or dam, thereby facilitating a conversion of the water's potential energy to kinetic energy. Accounting for over 16 percent of global electrical power generation in 2010, hydropower is the most widely used source of renewable energy. In 2010, hydropower accounted for 93 percent of the global renewable energy capacity. While the share of other renewable energy sources, particularly solar and wind, are expected to increase in the future, hydropower is projected to remain the largest renewable energy source for many years. The annual global hydroenergy potential is estimated to be over 16,400 TWh, but only about 19 percent of this potential has been developed.

Figure 4.1
The Hoover Dam, dedicated in 1935, has a power-generation capacity of 2080 MW.

(Timothy Oleary/ Shutterstock)

4.2 THE HYDROENERGY SOURCE

Kinetic energy of flowing water is the ultimate source of hydroenergy systems. Flowing water in rivers and streams is part of the water cycle of evaporation, condensation, precipitation, and runoff. Most of the earth's water is not available for use by hydroenergy systems. According to the U.S. Geological Survey (USGS), 96.54 percent of the earth's water is found in oceans, seas, and bays; 1.74 percent is found in ice caps, glaciers, and permanent snow; and 1.69 percent is in the form of groundwater. A much smaller percentage of the earth's water is found in ground ice, swamps, biological systems, and the atmosphere. Hence, flowing water in rivers and streams constitutes an extremely small percentage of the earth's water that can be exploited for hydroelectric use. Nevertheless, this extremely small percentage of water supplied the world over 3000 TWh of hydroelectric energy in 2010.

There are two methods by which electrical energy can be generated from flowing water. The first method is to convert the *kinetic* energy of water flowing in a river or stream to electrical energy. In the first system, known as a **run-of-river hydroelectric plant**, water flows through a turbine that converts the kinetic energy of the water into mechanical energy and a generator converts the mechanical energy to electrical energy. The fundamental principle of a run-of-river system is similar to that of a wind machine except that the fluid is water instead of air. The second method is to convert the *potential* energy of water that flows into a reservoir from a river or stream to electrical energy. In this type of system, known as a **conventional hydroelectric plant**, the potential energy of water stored in the reservoir is converted to kinetic energy by diverting the water through a conduit. A turbine in the conduit converts the kinetic energy of the water into mechanical energy, and a generator converts the mechanical energy to electrical energy. Both of these methods use flowing water to generate electrical energy, but the second, and most prevalent, method involves an intermediate stage where the water is first stored. The reason that this method is more prevalent is because the power generation of a run-of-river power plant is limited by the flow rate and breadth of the river. In a conventional power plant, the power generation is limited primarily by the depth of the water in the reservoir.

A variation of the conventional hydroelectric plant is the **pumped-storage plant**. In this type of system, water is moved between two reservoirs at different elevations. During times of low electrical power demand, excess generation capacity is used to pump water from the low reservoir to the high reservoir. During times of high demand, water is released back into the low reservoir through a turbine.

In the following section, we show how to calculate the available power in water in run-of-river and conventional hydroelectric plants.

4.2.1 Available Power in Water

Consider a run-of-river system, illustrated in Figure 4.2. From basic physics, kinetic energy is found using the formula

$$KE = \frac{1}{2} mv^2 \tag{4.1}$$

where m is mass and v is velocity. Dividing both sides of Equation (4.1) by mass, m, we obtain

$$ke = \frac{1}{2} v^2 \tag{4.2}$$

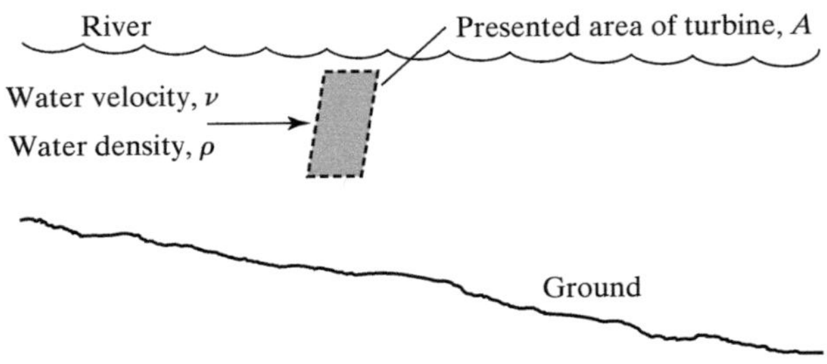

where ke is kinetic energy per unit mass, expressed in units of J/kg. The mass flow rate, $\dot{m}$, of water flowing through the presented area, A, of the turbine blades is expressed as

$$\dot{m} = \rho A v \tag{4.3}$$

where ρ is the density of water, expressed in units of kg/m^3, and $\dot{m}$ is expressed in units of kg/s. We see that by multiplying both sides of Equation (4.2) by mass flow rate, the left side of the equation now has units of J/s, which is defined as a watt (W), a unit of power. Thus, we have derived a relation for the available power in the flowing water of a run-of-river power plant,

$$P_{\text{water}} = \frac{1}{2}\rho A v^3 \tag{4.4}$$

Equation (4.4) is identical in form to Equation (3.5), the relation for available power in the wind, except that ρ is the density of water, not air; A is the presented area of the blades of a water turbine, not a wind turbine; and v is the velocity of water, not air. It should be noted that the shape of the presented area, A, depends on the type of water turbine used.

From Equation (4.4) we glean three important facts. First, available power in flowing water is proportional to water density. For water at standard conditions, $\rho = 1000$ kg/m^3. Second, available power is proportional to the presented area of the turbine blades. This applies to any type of water turbine. Third, as with available power in the wind, available power in flowing water is proportional to the cube of velocity.

Consider a conventional hydroelectric system, illustrated in Figure 4.3. From basic physics, gravitational potential energy is found using the formula

$$PE = mgz \tag{4.5}$$

where m is mass, g is gravitational acceleration (9.81 m/s^2), and z is height. Dividing both sides of Equation (4.5) by mass, m, we obtain

$$pe = gz \tag{4.6}$$

where pe is potential energy per unit mass, expressed in units of J/kg. Volume flow rate, $\dot{V}$, is the product of area and velocity, $\dot{V} = Av$, so Equation (4.3) can be expressed as

$$\dot{m} = \rho \dot{V} \tag{4.7}$$

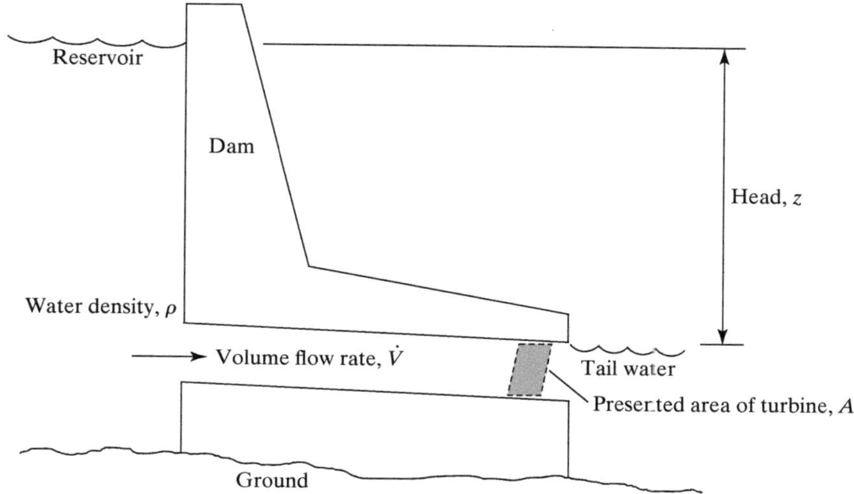

Figure 4.3
Diagram for calculating available waterpower for a conventional hydroelectric power plant.

We see that by multiplying both sides of Equation (4.6) by mass flow rate, the left side of the equation now has units of J/s, which is defined as a watt (W), a unit of power. Thus, we have derived a relation for the available power in the stored water of a conventional hydroelectric plant,

$$P_{\text{water}} = \rho \dot{V} gz \tag{4.8}$$

As shown in Figure (4.3), the quantity, z, referred to as the *head*, is the elevation difference between the surface of the water in the reservoir and the surface of the tail water. Volume flow rate, expressed in units of m^3/s, is the rate at which a given volume of water passes through the presented area, A, of the turbine blades. The calculation of volume flow rate requires an advanced knowledge of fluid dynamics, so volume flow rate will be treated here as a known quantity.

EXAMPLE 4.1

In a run-of-river hydroelectric plant, the velocity of the water is 4 m/s, and the presented area of the turbine blades is 10 m^2. Find the available power in the water.

SOLUTION

Using Equation (4.8), the available power in the water is

$$P_{\text{water}} = \frac{1}{2} \rho A v^3$$

$$= \frac{1}{2} (1000 \, \text{kg/m}^3)(10 \, \text{m}^2)(4 \, \text{m/s})^3$$

$$= 3.20 \times 10^5 \, \text{W} = 320 \, \text{kW}$$

EXAMPLE 4.2

In a conventional hydroelectric plant, the head is 40 m and the volume flow rate of water past the turbine blades is 1200 m^3/s. Find the available power in the water.

SOLUTION

Using Equation (4.4), the available power in the water is

$$P_{\text{water}} = \rho \dot{V} g z$$

$$= (1000 \, \text{kg/m}^3)(1200 \, \text{m}^3/\text{s})(9.81 \, \text{m/s}^2)(40 \, \text{m})$$

$$= 4.71 \times 10^8 \, \text{W} = 471 \, \text{MW}$$

PRACTICE!

1. The velocity of the water in a run-of-river hydroelectric plant is 6 m/s. If the presented area of the turbine blades is 14 m^2, find the available power in the water.

   Answer: 1.51 MW

2. In a small conventional hydroelectric plant, the head is 8 m and the volume flow rate of water through the turbine is 160 m^3/s. Find the available power in the water.

   Answer: 12.6 MW

4.3 HYDROPOWER SYSTEMS

Hydropower systems can be developed on a wide range of scales. Large conventional hydropower plants have power-generation capacities of 10 GW or more. The largest hydropower plant in the world is shown in Figure 4.4. The term **small hydro** refers to hydropower systems that power a small community or industrial facility. Small hydro systems, which have generation capacities of 10 MW or less, can be connected to the main power grid but are sometimes located in isolated regions and therefore power a limited area. A small hydro system can be a conventional power plant with a small reservoir or a run-of-river power plant with no reservoir. In a run-of-river system, water from the river or stream is diverted into a pipe or channel having a turbine, and the water is returned to the river downstream. Subclassifications of small hydro systems are known as *mini*, *micro*, and *pico* hydro, which have power-generation capacities of approximately 1000 kW, 100 kW, and 5 kW, respectively. In the next section, we consider large conventional hydropower and small hydropower systems further.

4.3.1 Large Conventional Hydropower

A conventional hydropower plant converts the potential energy of stored water to electrical energy. The potential energy of the water is proportional to the depth of the water in the reservoir. As illustrated in Figure 4.5, water enters a

Figure 4.4
The Three Gorges Dam, spanning the Yangtze River in China, is the world's largest hydropower plant with a power-generation capacity of 22.5 GW.

(Prill/Shutterstock)

downward-sloping channel called a *penstock*. A control gate at the entrance of the penstock is used to control the flow rate of water or close the penstock during periods of maintenance. A screen system at the intake prevents fish and debris from entering the penstock and potentially damaging the turbine. The turbine at the bottom of the penstock is fundamentally a water wheel that converts the kinetic energy of the flowing water to mechanical energy of a rotating shaft. The shaft of the turbine rotates the armature of the electrical generator, thereby producing electrical power, which is delivered to the power grid over transmission lines. Water exiting the turbine is carried by a draft tube to the tail water, which is typically a river.

Figure 4.5
Conventional hydropower plant.

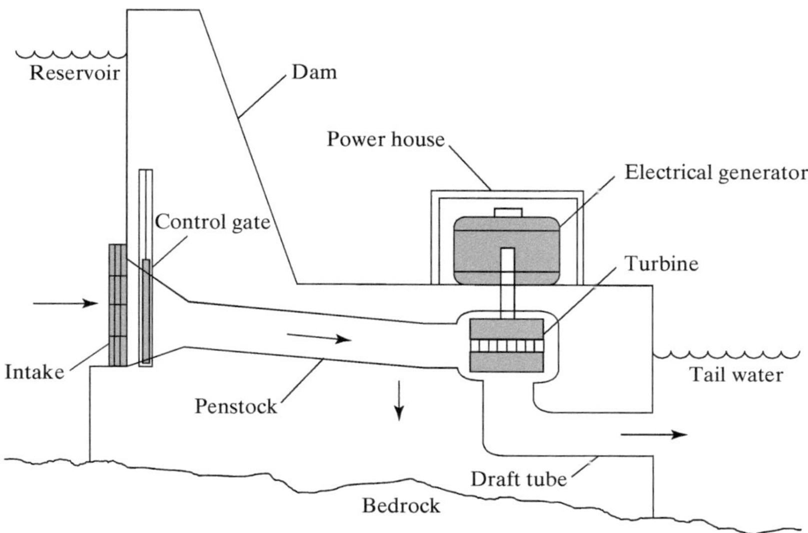

Like wind machines, the maximum electrical output power that a hydroelectric plant can generate is limited by energy losses in the system. Energy losses occur in the turbine and electrical generator, resulting in an electrical output power that is below the available power in the water given by Equation (4.8). Accounting for these losses, the electrical output power is given by the relation

$$P_{\text{elect}} = \eta P_{\text{water}} \qquad (4.9)$$

where P_{water} is given by Equation (4.8) and the quantity, η, is the **combined efficiency** of the turbine and electrical generator. For large hydroelectric plants, η ranges from about 0.85 to 0.95, whereas for small hydroelectric plants, η can be as low as 0.5.

The use of Equation (4.9) enables us to calculate the *instantaneous* electrical output power of a hydroelectric plant, that is, the electrical output power of a plant at a moment of time for a given head and volume flow rate. Of particular interest to engineers who design and analyze hydroelectric plants is the *total amount of electrical energy* that a hydroelectric plant generates during a specific time period. Because of seasonal changes that affect the reservoir on an annual cycle, this time period is typically one year. By finding the total electrical energy produced by a hydroelectric plant during a given time period, engineers can more fully assess the degree to which the plant meets the long-term electrical energy demands of a community.

The power plant **capacity factor**, *CF*, introduced in Chapter 1, is defined as the ratio of the actual electrical energy generated by the plant during a specific time period to the amount of electrical energy generated by the plant during the same time period, assuming the plant operates continuously at the rated capacity. Thus, the electrical energy generation of the plant is expressed as

$$E_{\text{elect}} = CF P_{\text{elect}} \Delta t \qquad (4.10)$$

where P_{elect}, which can be calculated using Equation (4.9), is considered the *rated capacity* or *installed capacity* of the plant, and Δt is the time period. As pointed out in Chapter 1, the time period is typically one year. Hydroelectric plant capacity factors vary widely, ranging from about 0.40 to 0.75, depending on the availability of water.

It is important to note that plant capacity factor can be used for calculating the electrical energy generation of any power plant, including a conventional fossil-fuel plant, a nuclear plant, or a renewable energy plant. The range of values of *CF* depends on the type of power plant.

EXAMPLE 4.3

The rated capacity of a large conventional hydroelectric plant is based on a head of 110 m and a volume flow rate of water of 3000 m³/s. The combined efficiency of the plant is 0.90, and the plant capacity factor is 0.65. Find the electrical energy generated by the plant during a one-year period.

SOLUTION

Using Equation (4.9), the electrical output power at rated capacity is

$$P_{\text{elect}} = \eta P_{\text{water}} = \eta \rho \dot{V} g z$$

$$= (0.90)(1000\,\text{kg/m}^3)(3000\,\text{m}^3/\text{s})(9.81\,\text{m/s}^2)(110\,\text{m})$$

$$= 2.91 \times 10^9\,\text{W}$$

This intermediate answer is expressed in units of J/s (W), so the time period, Δt, in Equation (4.10) is the number of seconds in a year,

$$\Delta t = 24\,\text{h/day} \times 3600\,\text{s/h} \times 365\,\text{day} = 3.15 \times 10^7\,\text{s}$$

Thus, the electrical energy generated by the plant in one year is

$$E_{\text{elect}} = CFP_{\text{elect}}\Delta t$$
$$= (0.65)(2.91 \times 10^9\,\text{J/s})(3.15 \times 10^7\,\text{s})$$
$$= 5.97 \times 10^{16}\,\text{J}$$

Our answer, expressed in the more commonly used electrical energy unit of kWh, is

$$E_{\text{elect}} = 5.97 \times 10^{16}\,\text{J} \times (1\,\text{kWh}/3.6 \times 10^6\,\text{J})$$
$$= 1.66 \times 10^{10}\,\text{kWh}$$

4.3.2 Small Hydropower

As previously stated, a small hydropower system can be a conventional power plant with a small reservoir or a run-of-river power plant with no reservoir. The basic energy analysis method presented in Section 4.3.1 for large hydropower systems also applies to conventional small hydropower systems. In this section, we show how to do a basic energy analysis of a small hydro run-of-river system.

Equation (4.4) is the relation for the available power in the water of a run-of-river system. Equations (4.9) and (4.10) for calculating the electrical output power and electrical output energy, respectively, of conventional hydroelectric plants also apply to run-of-river plants. In the following example, we analyze a pico hydro system.

EXAMPLE 4.4

Engineers working in a remote village with no central source of electricity utilize the power of a nearby stream to run a pico hydro system for powering lights in the village. At the peak of the spring runoff, the water flows at a velocity of 3.0 m/s. The turbine of the pico hydro system is a "water wheel" with a presented area of 0.75 m^2. If the combined efficiency of the system is 0.60, find the electrical output power during the spring runoff season.

SOLUTION

Combining Equations (4.4) and (4.9), the electrical output power of the pico hydro system is

$$P_{\text{elect}} = \frac{1}{2}\eta\rho A v^3$$

$$= \frac{1}{2}(0.60)(1000\,\text{kg/m}^3)(0.75\,\text{m}^2)(3.0\,\text{m/s})^3$$

$$= 6.08 \times 10^3\,\text{W} = 6.08\,\text{kW}$$

At spring runoff, this pico hydro system has the capacity to power one hundred 60 W light bulbs.

1. The reservoir of a large conventional hydroelectric plant has a head of 130 m and a volume flow rate of water of 4000 m^3/s. The combined efficiency of the plant is 0.94, and the plant capacity factor is 0.70. Find the electrical energy generated by the plant during a one-year period.

   Answer: 2.94×10^{10} kWh

2. A small hydropower system uses a reservoir with a water depth of 3.0 m. At capacity, the volume flow rate of water through the turbine is 5.0 m^3/s, and the combined efficiency of the system is 0.65. Should this system be categorized as a mini, micro, or pico hydro plant?

   Answer: micro hydro (P_{elect} = 95.6 kW)

4.3.3 Turbines

One of the key components of a hydroelectric plant of any size is the turbine. The purpose of the turbine is to convert kinetic energy of flowing water to mechanical energy of a rotating shaft. Modern turbines come in a variety of types and shapes, but the most commonly used turbine is the **Francis turbine**. The Francis turbine is a type of *reaction* turbine, a turbine in which the blades extract pressure energy from the water. A Francis turbine is typically mounted with the shaft oriented vertically, but because the turbine is completely submerged in the water entering from the penstock, the turbine can also be oriented horizontally. A typical Francis turbine wheel is shown in Figure 4.6.

The **Pelton turbine** operates on a completely different principle than the Francis turbine. A wheel with a set of double spoons or cups mounted around the rim, a Pelton turbine is driven by jets of high-velocity water impacting on each cup as the wheel rotates. Because energy from the water is imparted to the turbine via momentum transfer in short impulses, the Pelton turbine is referred to as an

Figure 4.6
Francis turbine wheel.

Figure 4.7
Pelton turbine wheel.

impulse turbine. In contrast to a reaction turbine where the turbine blades are completely submerged in water with a pressure difference across the blades, an impulse turbine operates in air at atmospheric pressure. A typical Pelton turbine wheel is shown in Figure 4.7.

PROFESSIONAL SUCCESS—PREPARING FOR A CAREER IN HYDROPOWER

Like the other renewable energy industries, the hydro industry employs engineers, scientists, technicians, and skilled workers from a variety of backgrounds. The primary engineering disciplines on which the hydroenergy industry relies are electrical, mechanical, civil, and environmental engineering. A hydroelectric facility is a complex electromechanical system consisting of turbines, pipes, pumps, and other mechanical components. Well-designed electrical generators and wiring systems are critical to the efficient generation and transmission of electrical power. Furthermore, a hydroelectric plant is a large civil engineering project that requires the careful design and construction of foundations, concrete structures, buildings, and roads. Hydroelectric plants, particularly large ones, can have a major impact on the environment, so environmental engineers are needed to study the environmental impacts of hydroelectric facilities and to help minimize them as much as possible. If the hydroelectric plant involves managing the reservoir and surrounding land, other nonengineering disciplines such as hydrology, biology, ecology, and recreation planners are also needed.

Students who major in any of the aforementioned disciplines have the potential academic preparation for a rewarding career in the hydropower industry.

4.4 ENVIRONMENTAL CONSIDERATIONS

Hydroenergy, like other renewable energy technologies, is environmentally benign in the respect that hydroelectric plants do not burn carbon-based fuels and therefore do not release greenhouse gases into the atmosphere. However, hydroelectric plants impact the environment in other significant ways. The primary environmental impacts are categorized as hydrological, ecological, geological, and social.

While a hydroelectric plant does not consume water, it alters the flow conditions at and downstream from the dam site. The natural environment determines flow conditions upstream of the reservoir, but the dam largely determines downstream flow conditions. Storing water in a reservoir reduces to some extent the flow downstream of the dam as a result of evaporation from a large exposed reservoir surface. The natural flow of water in a river or stream transports heavier-than-water particles downstream. A dam can disrupt this natural process by causing the gradual precipitation of sediment on the reservoir floor. Precipitation of sediment on the reservoir floor depletes sediment downstream of the dam, resulting in potential scouring of river beds and loss of river banks. Another potential hydrological effect is the alteration of ground water levels near the dam site.

Because a hydroelectric plant requires a reservoir for its operation, a large area of land is inundated, destroying plant life and destroying or displacing animal life. Furthermore, the natural habitat of the land surrounding the reservoir is affected, particularly if the reservoir is large. For hydroelectric plants along the Atlantic and Pacific coasts of North America, salmon populations have been reduced by limiting their access to spawning grounds upstream of the dams. Unless fish ladders are employed, fish that attempt to pass by the turbines are injured or killed. Furthermore, there are temperature gradients in deep bodies of water, so water exiting the turbines can be cooler or warmer than the water downstream depending on the depth at which the water is taken from the reservoir. These temperature differences are not large, but they can affect temperature-sensitive fauna that grows in river beds downstream of the dam.

The most potentially catastrophic impact of a hydroelectric plant is a structural failure due to a seismic event. Prior to the construction of a hydroelectric facility, a thorough study must be conducted to determine the geological suitability and stability of the site.

An obvious social impact of hydroelectric plants is the displacement of people who used to live in the region that was inundated by the reservoir. For example, over 1.2 million residents were displaced by the construction of the Three Rivers Dam in China. To many, however, a positive social impact of hydroelectric plants is the recreational benefit that the reservoir affords. The largest reservoir in the United States, Lake Mead, formed by Hoover Dam, provides visitors and local residents numerous recreational activities such as boating, water skiing, fishing, and swimming.

SUMMARY

Hydroenergy is the oldest renewable energy source, dating back to ancient Egypt and Mesopotamia. Electrical energy generation from hydroenergy became possible in the nineteenth century with the invention of the electrical generator. In 2010, hydroenergy accounted for over 16 percent of global electrical power generation and 93 percent of global renewable energy capacity.

There are two methods by which electrical energy can be generated by flowing water. The first method involves the conversion of kinetic energy of water flowing in a river or stream, and the second method involves the conversion of potential energy of water stored in a reservoir. The types of hydroelectric plants that employ these methods are referred to as run-of-river and conventional, respectively, of which conventional hydroelectric plants are the most common.

The available power in flowing water is calculated using the same mathematical relation that is used to calculate available power in wind. As with wind, available power in flowing water is proportional to the cube of velocity. Available power in stored water is proportional to water depth.

Hydroelectric plants have generation capacities ranging from less than 1 kW to over 10 GW. Small hydroelectric plants, categorized as either mini, micro, or pico hydro, have power-generation capacities of approximately 1000 kW, 100 kW, and 5 kW, respectively.

Combined efficiency of a hydroelectric plant is the product of turbine and electrical generator efficiencies. Plant capacity factor is the ratio of actual electrical energy generated by a hydroelectric plant during a period to the amount of electrical energy generated by the plant during the same period assuming the plant operates at rated capacity. This factor is typically used to calculate the annual amount of electrical energy generated by a hydroelectric plant.

There are two primary types of turbines in hydroelectric plants—reaction turbines and impulse turbines. A reaction turbine extracts pressure energy from flowing water, whereas an impulse turbine extracts energy via momentum transfer by impacting jets. The most commonly used reaction and impulse turbines are the Francis and Pelton turbine, respectively.

Hydroelectric plants do not release greenhouse gases into the atmosphere, but they affect the environment in other ways. Hydroelectric plants alter the flow and sedimentation conditions downstream of the dam. When a reservoir is constructed, a large amount of land is inundated, displacing plant and animal life and even human populations. Seismic events are the most potentially catastrophic impacts of hydroelectric facilities. Despite negative environmental impacts of hydropower, a positive social impact is the recreational use of reservoirs.

KEY TERMS

capacity factor	Francis turbine	pumped-storage plant
combined efficiency	hydroelectric	run-of-river hydroelectric
conventional	hydropower	plant
hydroelectric plant	Pelton turbine	small hydro

SUGGESTED READING

American Society of Mechanical Engineers, *The Guide to Hydropower Mechanical Design*, Kansas City, MO: HCI Publications, 1996.

Boyle, G., Ed., *Renewable Energy–Power for a Sustainable Future* 3rd Ed., Oxford: Oxford University Press, 2012.

Breeze, P., *Power Generation Technologies*, Burlington, MA: Elsevier, 2005.

Hagen, K.D., *Introduction to Engineering Analysis* 4th Ed., Upper Saddle River, NJ: Prentice Hall, 2014.

Khartchenko, N.V., and V.M. Khartchenko, *Advanced Energy Systems* 2nd Ed., Boca Raton, FL: CRC Press, 2014.

Kreith, F., *Principles of Sustainable Energy Systems* 2nd Ed., Boca Raton, FL: CRC Press, 2014.

Zooba, A.F., and R.C. Bansal, Eds., *Handbook of Renewable Energy Technology*, Hackensack, NJ: World Scientific, 2011.

PROBLEMS

For the following problems, it is recommended that you use the general analysis procedure of (1) problem statement, (2) diagram, (3) assumptions, (4) governing equations, (5) calculations, (6) solution check, and (7) discussion. This procedure is covered in Chapter 3 of Hagen, K.D., *Introduction to Engineering Analysis* 4th Ed.

The Hydroenergy Source

4.1 In a run-of-river hydroelectric power plant, the velocity of the water is 6 m/s, and the presented area of the turbine blades is 14 m^2. Find the available power in the water.

4.2 The equations for calculating available power in wind and flowing water are identical in form. Assuming the same turbine blade area and velocity for each system, how much more power is available in flowing water than wind?

4.3 In a run-of-river hydroelectric plant, find the required presented area of the turbine if the velocity of the water is 2.8 m/s. Use an available power of 90 kW.

4.4 In a conventional hydroelectric plant, the head is 70 m and the volume flow rate of water past the turbine blades is 3200 m^3/s. Find the available power in the water.

4.5 Construct a graph of the available power in flowing water by plotting available power as a function of water velocity. For velocity, use a range of 0 m/s to 12 m/s. For density of water and presented area of the turbine blades, use 1000 kg/m^3 and 1 m^2, respectively.

Large Conventional Hydropower

4.6 The rated capacity of a large conventional hydroelectric plant is based on a head of 130 m and a volume flow rate of water of 3800 m^3/s. The combined efficiency of the plant is 0.92, and the plant capacity factor is 0.70. Find the electrical energy generated by the plant during a one-year period.

4.7 At rated capacity, the head of a large conventional hydroelectric plant is 95 m, and the volume flow rate of water is 2200 m^3/s. The combined efficiency of the plant is 0.92, and the plant capacity factor is 0.61. Find the electrical energy generated by the plant during a one-year period.

4.8 The annual electrical energy requirement of a proposed hydroelectric plant is 3.0×10^{10} kWh. The combined efficiency and plant capacity factor are estimated to be 0.93 and 0.68, respectively. If the volume flow rate of water through the turbines is estimated to be 4400 m^3/s, what is the required head?

4.9 A hydroelectric plant has a gate at the spillway of the reservoir to regulate the flow of water downstream. The required range of electrical output power of the plant is 100 to 175 MW. If the head of the reservoir is held constant at 35 m, what is the range of flow rates that the gate must maintain if the combined efficiency is 0.90?

4.10 During a period of drought, a certain hydroelectric plant operates only 60 days of the year. The head and combined efficiency of the plant are 60 m and 0.89, respectively. For a volume flow rate of 1700 m^3/s, find the electrical energy generated by the plant during this one-year period.

4.11 From basic fluid dynamics, the volume flow rate of a liquid emanating from an opening in a container is given by the relation

$$\dot{V} = A\sqrt{2gz}$$

where A is the cross sectional area of the opening and z is elevation head (depth of the liquid with respect to the opening). See Figure P4.11. The primary restriction on the use of this relation is that there are no energy losses in the system, which is never strictly true for any system. Modeling the reservoir of a hydroelectric plant as a container of water, use this relation to estimate the electrical output power of the plant if the head is 85 m. The presented area of the turbine blades is 40 m^2, and the combined efficiency of the plant is 0.88.

Figure P4.11
Liquid emanating from a container.

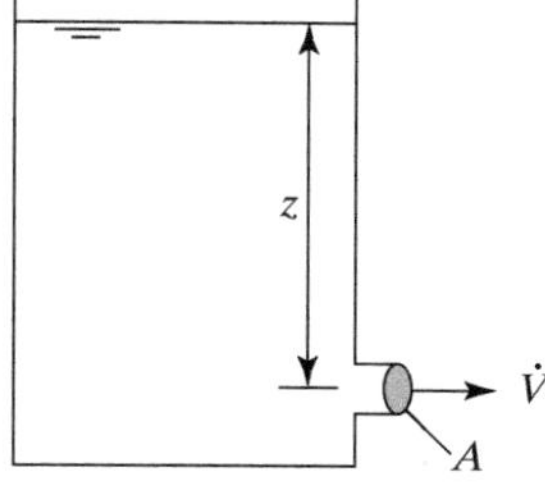

4.12 Find out the annual electrical energy demand of your city or town. How many Grand Coulee Dams would be required to meet this demand?

4.13 Write an essay on the economics of the Grand Coulee Dam. Include a discussion of the material and labor costs, governmental funding, and economic impact on the region.

Small Hydropower

4.14 A small hydroelectric plant uses a reservoir with a water depth of 4.5 m. The volume flow rate of water through the turbine is 80 m^3/s, and the combined efficiency of the system is 0.72. Calculate the electrical output power of the plant.

4.15 A run-of-river micro hydro plant employs a turbine with a presented blade area of 3.2 m^2. For a combined efficiency of 0.65, calculate the electrical output power of the plant for a water velocity of 2.8 m/s.

4.16 The annual electrical energy requirement of a building in a remote location is 3.0×10^5 kWh. A small run-of-river hydroelectric plant uses the energy of a nearby river to furnish electricity for the building. Assuming a combined efficiency and plant capacity factor of 0.81 and 0.62, respectively, what water velocity is required if the presented area of the turbine blades is 6.0 m^2?

4.17 In a pico hydro system for supplying electricity to a hut in a remote village, a square channel measuring 24 cm on a side diverts water from a stream. A turbine in the channel occupies three quarters of the channel cross section, and the combined efficiency of the turbine and generator is 0.30. Assuming the channel is full of water at all times, what is the electrical output power of the system if the velocity of the water is 3.8 m/s?

4.18 Pico hydro systems are becoming more popular as DIY (do it yourself) projects. Using material from this chapter and other sources as needed, conduct a paper design of a pico hydro system. If a hydroenergy source is available to you, build, install, and test the pico hydro system that you designed.

4.19 The initial cost of a DIY pico hydro system is $750. If the annual energy production of the system is 1400 kWh/y, and the price of electricity charged by the electric utility is $0.11/kWh, what is the simple payback?

Environmental Considerations

4.20 Write an essay on the environmental impacts of the Hoover Dam. In your essay, discuss the environmental issues before and during the construction of the dam. Also discuss the long-term environmental impacts of the Hoover Dam on the region.

4.21 The Three Gorges Dam in China is currently the largest hydroelectric plant in the world with a power-generation capacity of 22.5 GW. Write an essay on the environmental impacts of this facility, focusing your essay on the social aspects.

4.22 The Grand Coulee Dam, the largest hydroelectric plant in the United States, is located in the state of Washington on the Columbia River. Write an essay on the ecological and social environmental aspects of this facility.

4.23 Write an essay on the environmental advantages of micro and pico hydro systems. In your essay, discuss at least one operational micro or pico hydro plant, reporting on its location, power-generation capacity, and how the electrical power is used.

5 Geothermal Energy

Objectives

After reading this chapter, you will have learned:
- What geothermal energy is
- About the geothermal energy source
- How the different types of geothermal power plants work
- How to do a basic energy analysis of a geothermal power plant
- The environmental impacts of geothermal energy

5.1 INTRODUCTION

Geothermal energy is thermal energy generated and stored within the earth. Unlike other renewable energy sources, geothermal energy does not directly or indirectly originate from the sun. Approximately 80 percent of the earth's internal energy is heat generated by radioactive decay, and the remainder is residual heat from the formation of the planet. Heat from these two sources is transported to the earth's surface where it is manifested in hot ground and surface waters, rock, and magma. Even though the heat content of the earth is finite, geothermal energy is considered to be renewable because the rate of heat extraction from the earth is small compared with its heat content. The earth's thermal energy content is approximately 10^{31} J. This thermal energy is transported to the earth's surface at a rate of 44 TW, an amount of power that is more than double the global power consumption from all primary sources. However, most of this geothermal power cannot be exploited because of location.

Like hydroenergy, geothermal energy has been harnessed by humans for centuries. Hot springs from geothermal energy were used for bathing in the Paleolithic period in China and other regions. Geothermal energy was exploited later for space heating in ancient Rome. The first space heating systems in the United States using geothermal energy were developed in the nineteenth century. Homes in Iceland began using steam and hot water from geysers in the 1940s. Today, geothermal energy supplies space heating and hot water to nearly 90 percent of buildings in Iceland. During the twentieth century, the use of geothermal energy for generating electricity matured and continued to grow following the oil crisis of the 1970s. In 2010 geothermal energy accounted for approximately 10 GW of global electrical power generation. A projection suggests that geothermal energy could provide 1 percent of global electrical power generation by 2050.

5.2 THE GEOTHERMAL ENERGY SOURCE

Heat from radioactive decay and planetary accretion flows to the surface of the earth because of temperature gradients in the earth. The temperature of the center of the earth's core is approximately 7000°C, but the average temperature of the earth's surface is only 14°C. Because the earth's heat content is finite, the earth is not in thermal equilibrium, which means that heat flow to the surface is not steady. Thus, on a geological timescale, the earth is cooling. Throughout most of the earth's interior, heat is transported mainly by convection because the hot geological materials are in a deformable state. Closer to the earth's surface to a depth of approximately 100 km, heat is transported by conduction because the cooler geological materials are rigid.

A way of characterizing the geothermal energy source is to show the *heat flux* at the surface of the earth. Heat flux is defined as heat transfer rate per unit area, typically expressed in units of mW/m^2 (milliwatt/square meter). A map of the geothermal surface heat flux for the continental United States is shown in Figure 5.1. A region of high geothermal heat flux is indicative of high-temperature gradients and therefore means that the region is promising as a geothermal heat source. As shown on the map, the regions of highest geothermal heat flux are found in the western states. It is no wonder that the first geothermal district heating systems in the United States were built in Idaho and Oregon.

Figure 5.1 Geothermal surface heat flux map of the United States.

(Blackwell, D.D., Richards, M.C., Frone, Z.S., Batir, J.F., Williams, M.A., Ruzo, A.A., and Dingwall, R.K., 2011, "SMU Geothermal Laboratory Heat Flow Map of the Conterminous United States, 2011." Supported by Google.org. Available at http://www.smu.edu/geothermal.)

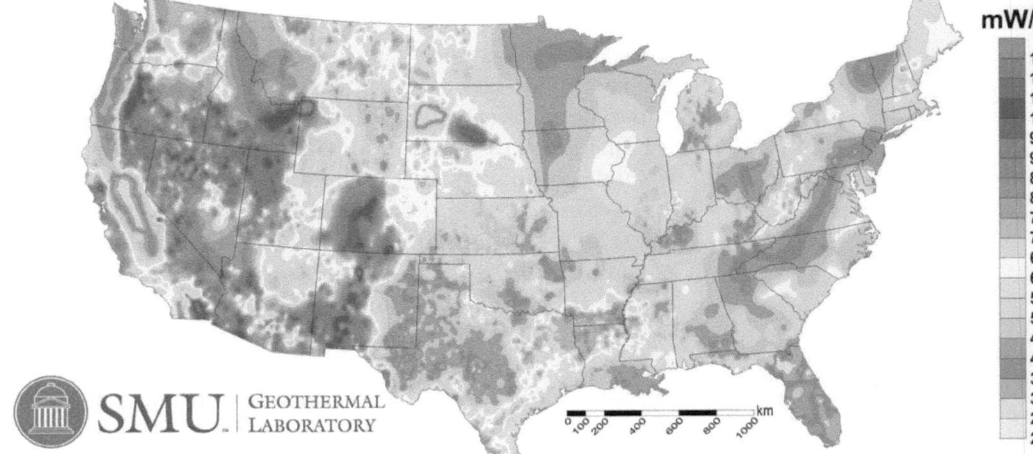

There are three primary geothermal energy sources. The first, and most readily implemented of the three, is hot underground water that either can be tapped by drilling boreholes or is naturally transported to the surface under pressure. The second is rock that is locally heated by natural formations in the earth's crust. This source is impractical to exploit because rock must be drilled. The third source is magma, which exists at very high temperatures and pressures, making the exploitation of this source very challenging.

Heat from a geothermal source can be converted to useful energy in three principal ways. First, underground hot water can be piped directly to heat living spaces of homes and businesses. A heat exchanger is used to transfer heat from the hot water to air that passes through the living space. This method is called **direct geothermal**. The second way utilizes a **heat pump**. Heat pumps take advantage of the year-round constancy of the earth's temperature a few meters below ground. Like a refrigerator, a heat pump incorporates a compressor and two heat exchangers. An underground heat exchanger transfers heat between the refrigerant and the ground, and a heat exchanger in the living space transfers heat between the refrigerant and indoor air. During the heating season, the earth is used as a heat source, whereas during the cooling season, the earth is used as a heat sink. This dual heating/air-conditioning function is accomplished by reversing the direction of the refrigerant. In the third way, steam or a mixture of steam and hot liquid water is used to generate electricity in a **geothermal power plant**.

There are three primary types of geothermal power plants. In the first type, the geothermal reservoir produces dry steam, much like the steam produced in the boiler of a conventional power plant, that is injected directly into a turbine. Dry steam means that the steam consists entirely of water vapor with no liquid water present. This type of plant is referred to as a **dry-steam power plant** or *direct-steam power plant*. A dry-steam geothermal power plant is shown in Figure 5.2. In the

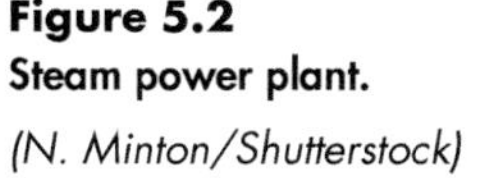

Figure 5.2
Steam power plant.

(N. Minton/Shutterstock)

second type, the geothermal reservoir produces a mixture of dry steam and liquid brine in which a portion of the liquid is converted to water vapor by a flash process. This type of plant is known as a **flash-steam power plant**. In the third type, the temperature of the geothermal reservoir is too low to produce steam, so heat from the reservoir is used to vaporize a secondary, low–boiling point fluid that is subsequently injected into the turbine. This type of plant is called a **binary power plant**.

5.3 GEOTHERMAL SYSTEMS

In this book we emphasize geothermal energy for the purpose of electrical power generation. The three geothermal power plants introduced earlier are discussed further in the following sections.

5.3.1 Dry-Steam Power Plant

A dry-steam power plant exploits a geothermal reservoir with a temperature range of about 180°C to 350°C. At these temperatures and the pressures supplied by the reservoir, the steam exists as a superheated vapor. Superheated steam is extracted from a production well in the reservoir and injected directly into a turbine. Because high-temperature, steam-rich geothermal reservoirs are rare, there are a limited number of dry-steam power plants in the world. Currently, there are only two such plants in the United States.

As illustrated in Figure 5.3, superheated steam is extracted from a borehole in a production well. Steam pressure is regulated using a throttling valve before the steam enters the turbine. Energy from the high-pressure, high-temperature steam is converted to mechanical energy of a rotating shaft in the turbine. The shaft of the turbine is connected to the shaft of the electrical generator. The electrical generator converts energy of its rotating shaft to electrical energy, which is transmitted across a power grid. Water exiting the turbine exists as a low-pressure saturated

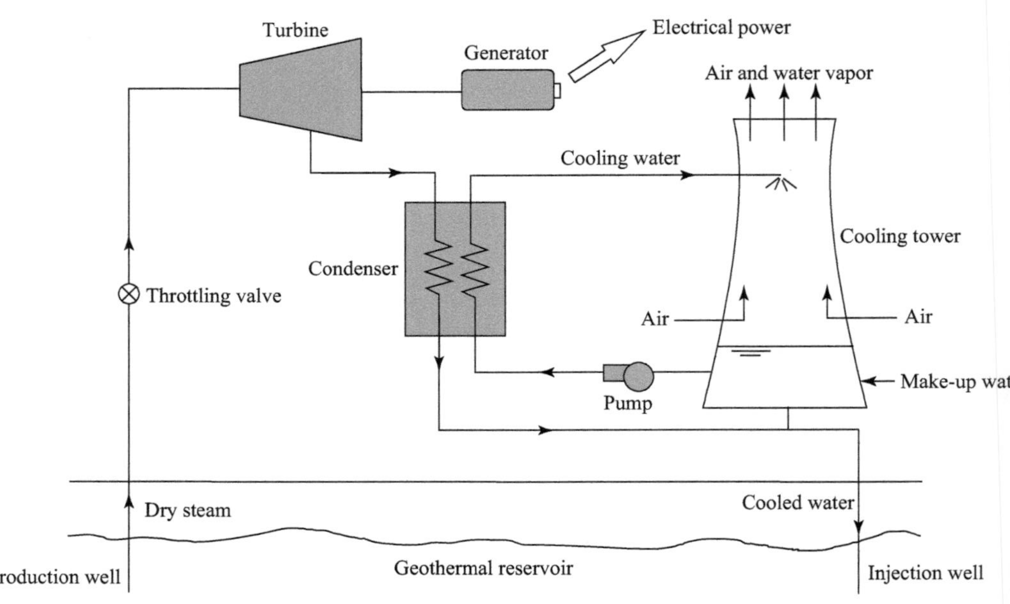

Figure 5.3
Diagram of a dry-steam geothermal power plant.

mixture of liquid and vapor that is condensed to a saturated liquid in a condenser that carries water from a cooling tower. The cooling tower transfers waste heat to the atmosphere by evaporation, and the cooled geothermal water is returned to the earth via an injection well.

5.3.2 Flash-Steam Power Plant

With the binary power plant, the flash-steam power plant is one of the most common types of geothermal power plants. In a flash-steam power plant, steam extracted from the geothermal reservoir consists of a mixture of steam and liquid brine (solution of salt and water). The temperature of the mixture in a flash-steam power plant is approximately 180°C or higher.

As illustrated in Figure 5.4, steam and brine are extracted from a borehole in a production well. Before entering the flash separator, the mixture is throttled by a throttling valve. The flash separator is a vessel maintained at a lower pressure than the mixture. The sudden reduction in pressure flashes (vaporizes) a portion of the hot mixture to water vapor. Gravity separates the heavier brine from the water vapor, which is piped to the turbine. Water exiting the turbine exists as a low-pressure saturated mixture of liquid and vapor that is condensed to a saturated liquid in a condenser carrying water from a cooling tower. The cooled geothermal water, along with the brine, is returned to the earth via an injection well.

A variation of the flash-steam power plant is the double-flash power plant in which the mixture from the first separator is flashed by a second separator. High-pressure steam from the first separator is sent to a primary turbine, and low-pressure steam from the second separator is sent to a secondary turbine. A double-flash power plant can generate up to 25 percent more power than a single-flash power plant, but double-flash power plants are more expensive and may be cost prohibitive.

Figure 5.4
Diagram of a flash-steam geothermal power plant.

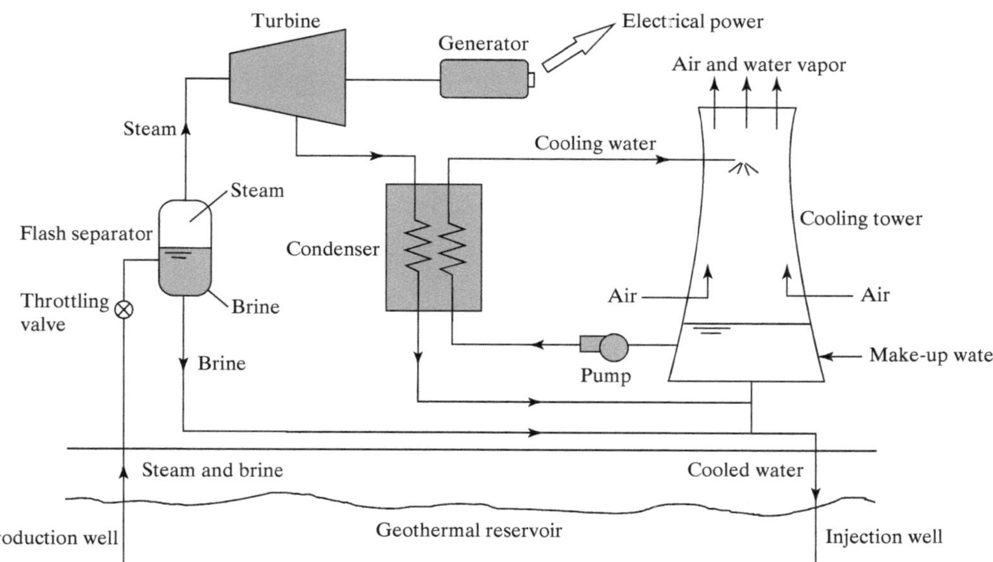

5.3.3 Binary Power Plant

If the steam from the geothermal reservoir is below approximately 180°C, the steam does not have enough thermal energy to efficiently power a turbine directly. However, energy can still be extracted from a low-temperature geothermal reservoir to generate electricity using a binary geothermal power plant. Because binary power plants operate at lower reservoir temperatures, this type of power plant can exploit more geothermal sites than dry-steam or flash-steam plants. For this reason, the binary power plant is the fastest growing type of geothermal power plant.

As illustrated in Figure 5.5, hot water from a production well flows through a heat exchanger where the hot water heats a low–boiling point working fluid such as isobutane. The cooled geothermal water is returned to the earth via an injection well. After exiting the turbine, the working fluid passes through a condenser where the vapor condenses to a liquid, exchanging heat with cooling water from a cooling tower. The working fluid circulates in a closed loop, never coming into physical contact with the geothermal water.

5.3.4 Basic Energy Analysis of a Geothermal Power Plant

A thorough energy analysis of a geothermal power plant, or a conventional power plant for that matter, requires a knowledge of thermodynamics and other engineering sciences that is beyond the scope of this book. For our purposes, a simplified but meaningful energy analysis is presented that pertains not only to a geothermal power plant but a conventional power plant as well.

A **heat engine** is a device that converts heat to work, and this is precisely what a power plant does. This definition applies to both geothermal power plants and conventional power plants. A power plant converts heat to shaft work of a turbine, which is converted to electrical energy. As illustrated in Figure 5.6, a heat engine receives an amount of heat, Q_{in}, from a high-temperature source and converts a

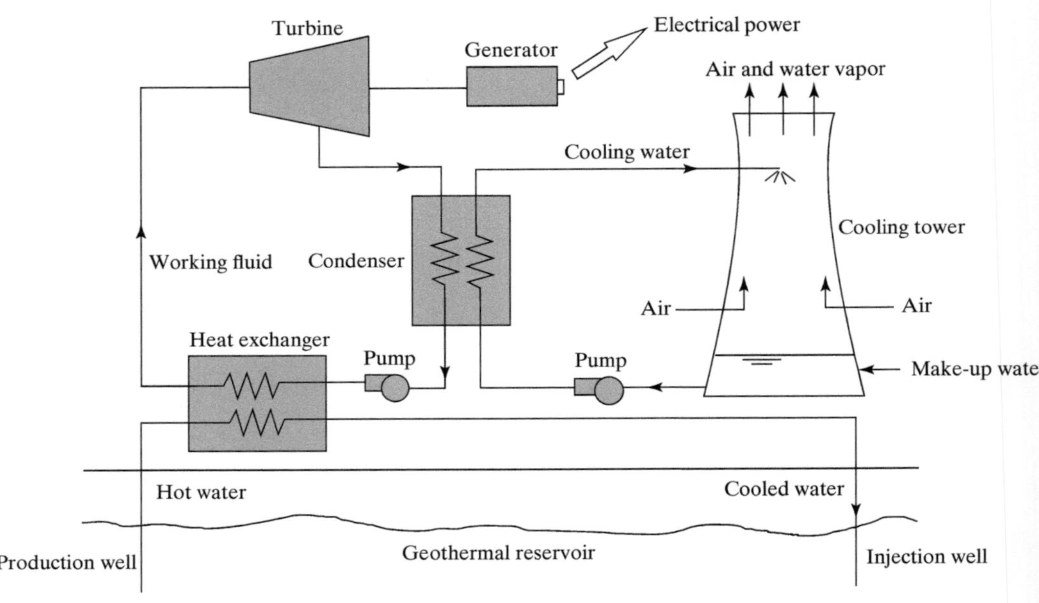

Figure 5.5
Diagram of a binary geothermal power plant.

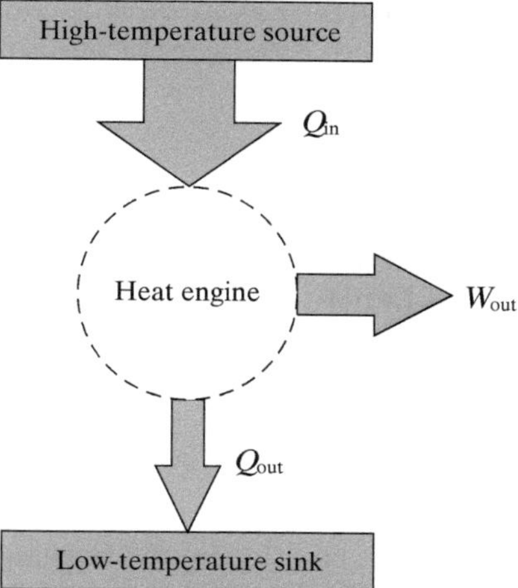

portion of that heat to work, W_{out}. The heat engine rejects the remaining heat, Q_{out}, to a low-temperature sink. In a conventional power plant, Q_{in} is the heat supplied to the boiler from burning a fossil fuel such as coal, oil, or natural gas. In the case of a nuclear power plant, heat is supplied by the decay of a radioactive material. In a geothermal power plant, Q_{in} is the heat supplied by a geothermal source. In all types of power plants, Q_{out} is heat rejected to the environment. Typically, Q_{out} is heat rejected to the atmosphere via a cooling tower, but Q_{out} can also be heat rejected to a natural body of water, such as an ocean, lake, or river, via a heat exchanger. The quantity W_{out} is the shaft work (energy) generated by the turbine.

The **first law of thermodynamics** states that *energy is conserved.* Thus, for a heat engine, the heat supplied must equal the output work plus the rejected heat. Expressed mathematically, the first law of thermodynamics for the heat engine is written as

$$Q_{in} = Q_{out} + W_{out} \tag{5.1}$$

where each quantity is expressed in units of the joule (J). Stated in terms of power, these quantities can also be expressed in units of the watt (W). **Thermal efficiency** of a heat engine, denoted η_{th}, is defined as the shaft work of the turbine divided by the heat input,

$$\eta_{th} = \frac{W_{out}}{Q_{in}} \tag{5.2}$$

The **second law of thermodynamics** states that *it is impossible for a heat engine to convert all the heat it receives from a high-temperature source to work.* Hence, the value of η_{th} is limited to values less than 1. A well-known theorem from thermodynamics states that for an *ideal* heat engine operating between source and sink temperatures of T_H and T_L, respectively, the maximum thermal efficiency is given by the relation

$$\eta_{th,ideal} = 1 - \frac{T_L}{T_H} \tag{5.3}$$

where T_H and T_L are expressed in absolute temperature units of kelvin (K) or rankine (R). The thermal efficiency given by Equation (5.3) is the maximum possible thermal efficiency a heat engine can have and is referred to as the **Carnot efficiency**, in honor of the French engineer Sadi Carnot.

Turbine shaft work is converted to electrical energy by the generator, which means that the generator also has an efficiency. Generator efficiency is defined as electrical energy divided by shaft work,

$$\eta_{\text{gen}} = \frac{E_{\text{elect}}}{W_{\text{out}}} \tag{5.4}$$

Generator efficiencies typically exceed 0.90. The quantities on the right-hand side of Equation (5.4) can also be expressed in terms of power. The product of the thermal and generator efficiencies given by Equations (5.2) and (5.4) defines the overall efficiency, η_{overall},

$$\eta_{\text{overall}} = \eta_{\text{th}}\,\eta_{\text{gen}} = \frac{W_{\text{out}}}{Q_{\text{in}}}\frac{E_{\text{elect}}}{W_{\text{out}}} = \frac{E_{\text{elect}}}{Q_{\text{in}}} \tag{5.5}$$

where, once again, the heat and work quantities can be expressed in terms of power.

The use of Equation (5.5) enables us to calculate the *instantaneous* electrical output power of a geothermal power plant, that is, the electrical output power of a plant at a moment of time for a given set of conditions. Of particular interest to engineers who design and analyze geothermal power plants is the *total amount of electrical energy* that a geothermal plant generates during a specific time period. Unlike solar, wind, and hydro plants that are affected by weather and seasonal conditions, thermal conditions in the earth are essentially unaffected by these factors. Thus, geothermal power plants can provide nearly constant power throughout the year.

The power plant **capacity factor**, *CF*, introduced in Chapter 1, is defined as the ratio of the actual electrical energy generated by the plant during a specific time period to the amount of electrical energy generated by the plant during the same time period, assuming the plant operates continuously at the rated capacity. Since power is defined as energy divided by time, the electrical energy generation of the plant may be expressed as

$$E_{\text{elect}} = CF\,P_{\text{elect}}\,\Delta t \tag{5.6}$$

where P_{elect} is electrical output power and Δt is the time period. Geothermal plant capacity factors are typically very high, ranging from about 0.85 to 0.90.

It is important to note that plant capacity factor can be used for calculating the electrical energy generation of any power plant, including a conventional fossil-fuel plant, a nuclear plant, or a renewable energy plant. The range of values of *CF* depends on the type of power plant.

In the following three examples, we illustrate the use of the earlier equations for a dry-steam power plant, a flash-steam power plant, and a binary power plant.

EXAMPLE 5.1

In a dry-steam power plant, 325°C superheated steam from the production well supplies 12.0 MW to the turbine. The output power of the turbine shaft is 4.2 MW, and the efficiency of the electrical generator is 0.94. Find the thermal efficiency, overall efficiency, electrical output power, rate of heat rejection to a 20°C atmosphere, and Carnot efficiency. If the plant capacity factor is 0.90, how much electrical energy does the plant generate in one year?

SOLUTION

From the power version of Equation (5.2), thermal efficiency is

$$\eta_{th} = \frac{\dot{W}_{out}}{\dot{Q}_{in}}$$

$$= \frac{4.2 \text{ MW}}{12 \text{ MW}}$$

$$= 0.350$$

Using Equation (5.5), overall efficiency is

$$\eta_{overall} = \eta_{th}\eta_{gen}$$

$$= (0.350)(0.94)$$

$$= 0.329$$

Using the power version of Equation (5.5), the electrical output power is

$$P_{elect} = \eta_{overall}\dot{Q}_{in}$$

$$= (0.329)(12.0 \text{ MW})$$

$$= 3.95 \text{ MW}$$

Using Equation (5.1), the first law of thermodynamics for a heat engine, the rate of heat rejection to the atmosphere is

$$\dot{Q}_{out} = \dot{Q}_{in} - \dot{W}_{out}$$

$$= 12 \text{ MW} - 4.2 \text{ MW}$$

$$= 7.80 \text{ MW}$$

Converting the geothermal steam and atmosphere temperatures to units of kelvin, the Carnot efficiency is

$$\eta_{th,ideal} = 1 - \frac{T_L}{T_H}$$

$$= 1 - \frac{(20 + 273)\text{K}}{(325 + 273)\text{K}}$$

$$= 0.510$$

The time period, Δt, in Equation (5.6) is the number of seconds in a year,

$$\Delta t = 24 \text{ h/day} \times 3600 \text{ s/h} \times 365 \text{ day} = 3.15 \times 10^7 \text{ s}$$

Thus, the electrical energy generated by the plant in one year is

$$E_{elect} = CF\, P_{elect}\, \Delta t$$

$$= (0.90)(3.95 \times 10^6 \text{ J/s})(3.15 \times 10^7 \text{ s})$$

$$= 1.12 \times 10^{14} \text{ J}$$

Our answer, expressed in the more commonly used electrical energy unit of kWh, is

$$E_{elect} = 1.12 \times 10^{14} \text{ J} \times (1 \text{ kWh}/3.6 \times 10^6 \text{ J})$$

$$= 3.11 \times 10^7 \text{kWh}$$

(continued)

Our calculations show that $\eta_{th} < \eta_{th,ideal}$. This result is demanded by the second law of thermodynamics, a version of which states that the actual thermal efficiency of a heat engine cannot exceed that of a Carnot heat engine operating between the same source and sink temperatures.

EXAMPLE 5.2

The production well of a flash-steam power plant supplies 4.5 MW to the steam entering the flash separator, and the thermal and generator efficiencies are 0.28 and 0.89, respectively. The temperature of the steam at the production well is 175°C, and the temperature of the atmosphere is 20°C. Find the electrical output power, rate of heat rejection to the atmosphere, and Carnot efficiency.

SOLUTION

From Equation (5.5), the overall efficiency is

$$\eta_{overall} = \eta_{th}\,\eta_{gen}$$
$$= (0.28)(0.89)$$
$$= 0.249$$

Thus, the electrical output power is

$$P_{elect} = \eta_{overall}\dot{Q}_{in}$$
$$= (0.249)(4.5\ \text{MW})$$
$$= 1.12\ \text{MW}$$

Using Equation (5.2), the turbine output power is

$$W_{out} = \eta_{th}\dot{Q}_{in}$$
$$= (0.28)(4.5\ \text{MW})$$
$$= 1.26\ \text{MW}$$

so, using Equation (5.1), the rate of heat rejection to the atmosphere is

$$\dot{Q}_{out} = \dot{Q}_{in} - \dot{W}_{out}$$
$$= 4.5\ \text{MW} - 1.26\ \text{MW}$$
$$= 3.24\ \text{MW}$$

Using Equation (5.3), the Carnot efficiency is

$$\eta_{th,ideal} = 1 - \frac{T_L}{T_H}$$
$$= 1 - \frac{(20 + 273)\text{K}}{(175 + 273)\text{K}}$$
$$= 0.346$$

EXAMPLE 5.3

In a binary power plant, 55°C water from the production well supplies 400 kW to an isobutane working fluid in the heat exchanger. The output power of the turbine shaft is 30 kW, and the efficiency of the generator is 0.90. Find the thermal efficiency, overall efficiency, electrical output power, and rate of heat rejection to the atmosphere. If the temperature of the atmosphere is 18°C, find the Carnot efficiency.

SOLUTION

From the power version of Equation (5.2), thermal efficiency is

$$\eta_{th} = \frac{\dot{W}_{out}}{\dot{Q}_{in}}$$

$$= \frac{30 \text{ kW}}{400 \text{ kW}}$$

$$= 0.075$$

From Equation (5.5), overall efficiency is

$$\eta_{overall} = \eta_{th}\eta_{gen}$$

$$= (0.075)(0.90)$$

$$= 0.0675$$

From Equation (5.5), the electrical output power is

$$P_{elect} = \eta_{overall}\dot{Q}_{in}$$

$$= (0.0675)(400 \text{ kW})$$

$$= 27.0 \text{ kW}$$

Using Equation (5.1), the rate of heat rejection to the atmosphere is

$$\dot{Q}_{out} = \dot{Q}_{in} - \dot{W}_{out}$$

$$= 400 \text{ kW} - 30 \text{ kW}$$

$$= 370 \text{ kW}$$

Converting the geothermal water and atmosphere temperatures to units of kelvin, the Carnot efficiency is

$$\eta_{th,ideal} = 1 - \frac{T_L}{T_H}$$

$$= 1 - \frac{(18 + 273)\text{K}}{(55 + 273)\text{K}}$$

$$= 0.113$$

Thermal efficiencies of binary power plants are typically low, so our calculated value of 0.0675 is not surprising. Note that the actual thermal efficiency is lower than the Carnot efficiency, as required by the second law of thermodynamics.

PRACTICE!

1. A dry-steam geothermal power plant operates on 300°C steam. For an atmosphere temperature of 17°C, find the maximum possible thermal efficiency of the power plant.

   Answer: 0.494

2. The production well of a dry-steam geothermal power plant supplies 25.0 MW to the steam that enters the turbine. The shaft of the turbine produces 6.2 MW of mechanical power, and the efficiency of the electrical generator is 0.95. Find the thermal efficiency, overall efficiency, electrical output power, and rate of heat rejection to the atmosphere.

   Answer: 0.248, 0.236, 5.89 MW, 18.8 MW

3. The thermal and electrical generator efficiencies of a binary power plant are 0.43 and 0.88, respectively. If the electrical output power of the plant is 2.8 MW, find the rate of heat transfer from the hot water supplied by the production well.

   Answer: 7.40 MW

PROFESSIONAL SUCCESS—ELECTRICAL POWER GENERATION: THE BIG PICTURE

In the United States, fossil fuels are the primary sources of electrical power. According to the U.S. Energy Information Administration (EIA), fossil fuels (coal, natural gas, and petroleum) accounted for 68 percent of electrical power generation in 2013, while renewable sources accounted for 13 percent and nuclear accounted for 19 percent. Of the renewable sources, hydro accounted for 52 percent, wind accounted for 31 percent, biomass accounted for 12 percent, geothermal accounted for 3 percent, and solar accounted for 2 percent.

Most hydroelectric power plants in the United States were built before 1975, and federal agencies operate most of the dams. Electrical power generated by wind turbines has increased significantly in the past decade. This increase is primarily due to federal financial incentives and renewable energy standards mandated by the states. Solar systems generated a significant amount of electrical power at small installations such as residential rooftops. The majority of biomass is municipal solid waste that is burned to provide electrical power.

The United States is second in the world in electrical power generation from renewable sources. China is the world leader in electrical power generation from renewable sources primarily because of recent additions to hydroelectric power capacity.

The question is often asked, "Why don't we use more renewable energy to generate electricity?" There is one central reason—cost. Renewable energy power plants are typically more expensive to build and operate per unit of energy produced than conventional power plants. Furthermore, some renewable energy sources are located in remote regions, and building transmission lines from the plants to the users of the energy is expensive.

Specific energy policies in the United States are intended to promote renewable energy use by mitigating costs. Tax credits, such as the Renewable Electricity Production Tax Credit, are federal incentives to boost the implementation of wind and other renewable energy sources. Many states have implemented standards that require electricity providers to generate or acquire a certain portion of their electrical power from renewable sources. Also, some states have built renewable energy credits into their standards that allow electricity providers to sell those credits.

Fossil fuels will most likely be the predominant sources of electrical power in the near future, but renewable energy sources are gaining ground. As long as cost incentives are preserved, or even enhanced, and care for the environment is sustained, this trend will continue.

5.4 ENVIRONMENTAL CONSIDERATIONS

Compared to fossil-fuel power plants, geothermal power plants are environmentally benign. However, geothermal power plants are not without environmental impacts. As water is carried from the earth, some dissolved gases are carried with it, notably carbon dioxide (CO_2), methane (CH_4), hydrogen sulfide (H_2S), and ammonia (NH_3). When introduced into the atmosphere, these gases contribute to climate change and acid rain and, with the exception of CO_2, emit unpleasant odors. As a fraction of generated electrical power, geothermal power plants produce much less of these pollutants than fossil-fuel plants. Geothermal water can also contain trace amounts of toxic elements such as mercury and arsenic. The release of harmful substances into the environment is mitigated by injecting the geothermal fluids back into the earth via the injection well. Continuous operation of a geothermal power plant may result in lowering the local water table, but reinjection of the water mitigates this problem as well. Reinjection can cause a local cooling of the reservoir, but this effect can be minimized by mapping the local reservoir flows and reinjecting the cooled water a suitable distance away from the production well.

Ground subsidence (settling) and seismic activity have been attributed to the long-term operation of geothermal power plants. The magnitude of ground subsidence is typically less than a centimeter, and the link to earthquakes is questionable because geothermal sites are generally prone to earthquakes anyway.

The environmental impact of the construction of geothermal power plants is comparable to that of conventional plants. Unlike conventional plants, however, geothermal plants involve the drilling of wells, which requires great quantities of water from the surroundings. Geothermal drilling is safer than oil or gas drilling, however, because there is no risk of fire, and the potential environmental impact of a drilling accident is relatively minor. Lastly, geothermal power plants require small areas of land, taking up only a few acres per 100 MW of power generation.

SUMMARY

Geothermal energy is thermal energy generated and stored within the earth. Approximately 80 percent of geothermal energy is caused by radioactivity, and the remainder is residual heat from planetary accretion. The heat content of the earth is finite, but geothermal energy is considered to be renewable because the rate of heat extraction from the earth is small compared with its heat content, which is approximately 10^{31} J. Geothermal energy has been used by humans for centuries for space heating and bathing, but for electrical power generation, geothermal energy has its beginning in the twentieth century. In 2010 geothermal energy accounted for about 10 GW of global electrical power generation.

Heat flows to the earth's surface due to temperature gradients in the earth. The temperature difference between the earth's core and its surface is approximately 7000°C. There are three main geothermal energy sources: hot underground water, hot rock, and magma. Energy from these sources can be used to directly heat living spaces, to run a heat pump, or to generate electricity in a power plant.

The three primary types of geothermal power plants are dry steam, flash steam, and binary. In a dry-steam power plant, superheated steam from a production well is injected directly into a turbine. In a flash-steam power plant, steam is separated from a mixture of steam and brine in a separator where the steam is delivered to the turbine. In a binary power plant, hot water transfers heat to a low–boiling point fluid that is piped to the turbine. A basic energy analysis of these three power plants can be done by employing the first and second laws of thermodynamics and efficiency relations for the turbine and electrical generator.

Compared to conventional power plants, geothermal power plants are environmentally friendly. A small amount of harmful gases is released into the atmosphere, but this can be mitigated by injecting the geothermal fluids back into the earth via the injection well. Reinjection also mitigates the lowering of the water table and localized cooling of the geothermal reservoir. Ground settling and seismic activity have been associated with geothermal power plants. The environmental impact of the construction of geothermal power plants is comparable to that of conventional power plants.

KEY TERMS

binary power plant	first law of	heat engine
capacity factor	thermodynamics	heat pump
Carnot efficiency	flash-steam power	second law of
direct geothermal	plant	thermodynamics
dry-steam power plant	geothermal power plant	thermal efficiency

SUGGESTED READING

BOYLE, G., Ed., *Renewable Energy–Power for a Sustainable Future* 3rd Ed, Oxford: Oxford University Press, 2012.

BREEZE, P., *Power Generation Technologies*, Burlington, MA: Elsevier, 2005.

HAGEN, K.D., *Introduction to Engineering Analysis* 4th Ed, Upper Saddle River, NJ: Prentice Hall, 2014.

KHARTCHENKO, N.V. and V.M. KHARTCHENKO, *Advanced Energy Systems* 2nd Ed., Boca Raton, FL: CRC Press, 2014.

ZOOBA, A.F. and R.C. BANSAL, Eds., *Handbook of Renewable Energy Technology*, Hackensack, NJ: World Scientific, 2011.

PROBLEMS

For the following problems, it is recommended that you use the general analysis procedure of (1) problem statement, (2) diagram, (3) assumptions, (4) governing equations, (5) calculations, (6) solution check, and (7) discussion. This procedure is covered in Chapter 3 of Hagen, K.D., *Introduction to Engineering Analysis* 4th Ed.

The Geothermal Energy Source

5.1 The heat content (internal energy) of a solid can be calculated using the relation

$$U = m\,c\,T$$

where U is internal energy in units of J, m is mass in units of kg, c is specific heat in units of J/kg·°C, and T is temperature in units of °C. The specific heat of granite is approximately 800 J/kg·°C. Find the internal energy per unit mass of 200°C granite.

5.2 The relation in Problem 5.1 also applies to liquids. Find the internal energy per unit mass of 190°C liquid water. Use a specific heat of 4875 J/kg·°C.

5.3 A well-known relation in heat transfer is Fourier's law of heat conduction

$$q'' = k\Delta T/\Delta x$$

where q'' is heat flux in units of W/m^2, k is thermal conductivity in units of W/m·°C, and $\Delta T/\Delta x$ is temperature gradient (temperature change divided by distance) in units of °C/m. The thermal conductivity of granite is approximately 2.8 W/m·°C. Using this value for k in Fourier's law, find the heat flux at the earth's surface if the temperature difference across the earth's crust to a depth of 7000 m is 300°C. Is the answer consistent with the heat flux data given in Figure 5.1?

Geothermal Systems

5.4 A dry-steam power plant operates between the temperature limits of 15°C and 380°C. Find the maximum thermal efficiency of this power plant.

5.5 The design engineers of a binary power plant use a maximum theoretical thermal efficiency of 0.16 in the preliminary phase of the design. If the engineers assume that the temperature of the atmosphere is 20°C, what is the required temperature of the geothermal heat source?

5.6 The steam from the production well of a dry-steam power plant supplies 8.5 MW to a turbine. If the output work of the turbine is 3.9 MW, what is the thermal efficiency of the power plant?

5.7 The thermal and electrical generator efficiencies of a flash-steam power plant are 0.55 and 0.92, respectively. If the heat input from the steam in the production well is 16 MW, what is the electrical output power of the plant?

5.8 The production well of a dry-steam power plant supplies 30 MW to the steam that enters the turbine. The efficiency of the electrical generator is 0.90, and the shaft of the turbine produces 9.5 MW of mechanical power. Find the thermal efficiency, overall efficiency, electrical output power, and rate of heat rejection to the atmosphere.

5.9 The thermal and electrical generator efficiencies of a binary power plant are 0.40 and 0.90, respectively. If the electrical output power of the plant is 3.6 MW, find the rate of heat transfer from the hot water at the production well.

5.10 In a dry-steam power plant, 325°C superheated steam from the production well supplies 16.8 MW to the turbine. The output power of the turbine shaft is 7.0 MW, and the efficiency of the electrical generator is 0.94. Find the thermal efficiency, overall efficiency, electrical output power, rate of heat rejection to a 20°C atmosphere, and Carnot efficiency.

5.11 If the plant capacity factor for the dry-steam power plant in Problem 5.10 is 0.85, how much energy does the plant generate in one year?

5.12 In a binary power plant, 70°C water from the production well supplies 350 kW to an isobutane working fluid in the heat exchanger. The output power of the turbine shaft is 25 kW, and the efficiency of the generator is 0.95. Find the thermal efficiency, overall efficiency, electrical output power, and rate of heat rejection to the atmosphere. If the temperature of the atmosphere is 22°C, find the Carnot efficiency.

5.13 Consider the following parameters for a dry-steam power plant. Steam at 280°C at the production well supplies 30 MW of heat. The output power of the turbine shaft is 16 MW, and the temperature of the atmosphere is 20°C. By analysis, show that this power plant is impossible.

5.14 In a dry-steam power plant, 325°C superheated steam from the production well supplies 20 MW to the turbine. The output power of the turbine shaft is 6.5 MW, and the efficiency of the electrical generator is 0.91. Find the thermal efficiency, overall efficiency, electrical output power, rate of heat rejection to a 10°C atmosphere, and Carnot efficiency. If the plant capacity factor is 0.88, how much electrical energy does the plant generate in one year?

Environmental Considerations

5.15 The Geysers north of San Francisco, California, is a complex consisting of 22 geothermal power plants. Write an essay on the environmental impacts of this facility. In your essay, discuss the environmental issues before and during the construction of the geothermal field. Also discuss the long-term environmental impacts of the facility on the region.

5.16 Write an essay on geothermal energy in Iceland. Discuss the primary uses of geothermal energy on the island and how the predominance of this renewable energy form has impacted the region.

6 Marine Energy

Objectives

After reading this chapter, you will have learned:

- What marine energy is
- About the marine energy sources
- How the different types of marine power plants work
- How to do a basic energy analysis of a marine power plant
- The environmental impacts of marine energy

6.1 INTRODUCTION

Marine energy is a comprehensive term used to describe energy derived from oceans. There are two main types of marine energy. The first type is **tidal energy**, which is generated by the periodic rise and fall of sea levels. Tides are primarily caused by gravitational forces exerted by the moon and sun but are also caused by the earth's rotation. The second type, called **ocean energy**, can be broken into three subtypes that exploit different physical phenomena: **ocean currents**, **ocean waves**, and **ocean thermal energy**. In the first two subtypes, energy in ocean currents and waves is converted to mechanical energy, which is subsequently converted to electrical energy. (Ocean currents are actually a consequence of tides, so this type of ocean energy can be classified as a type of tidal energy.) In the third subtype, a heat engine utilizes temperature differences between deep and shallow ocean water to generate mechanical energy in a turbine and then electrical energy in a generator. This type of system is called **ocean thermal energy conversion** (OTEC).

The history of tidal motion to produce mechanical power extends back to the Middle Ages. Tidal basins were used in Europe to power grist mills prior to 1000 A.D. This technology continued until the Industrial Revolution of the nineteenth century when steam power replaced it. The use of tidal motion for generating electrical power is relatively new, dating back only to the mid-twentieth century.

It is estimated that the global energy generation of tidal motion is approximately 22,000 TWh. Of this amount, only 200 TWh can be economically utilized, and less than 0.6 TWh of tidal energy is converted to electrical energy. The only tidal power plant currently in operation in North America is the Annapolis Royal Generating Station located on the Annapolis River just upstream from the city of Annapolis Royal, Nova Scotia, Canada (see Figure 6.1). The La Rance plant is a tidal barrage facility with a power-generation capacity of approximately 240 MW. The first commercially licensed tidal power plant in the United States is the Roosevelt Island Tidal Energy (RITE) project located in the strait that connects the Long Island Sound with the Atlantic Ocean in New York Harbor. By 2015 this power plant will have 30 turbines in the strait that are powered by currents caused by tidal motion. The power-generation capacity of the RITE plant is approximately 1 MW. One of the turbines for this power plant just prior to installation is shown in Figure 6.2.

The global energy-generation capacity of ocean currents is estimated to be 2000 TWh, and the global energy capacity of ocean waves is estimated to be between 1 TWh and 10 TWh. Unlike tidal energy, ocean energy has only recently been exploited. Following the oil crisis of the 1970s, the use of ocean currents to generate electrical power became an active area of renewable energy research. In the years that followed, several experimental programs demonstrated that ocean currents, waves, and temperature gradients have the potential for generating a significant amount of electrical power.

The world's first wave energy farm was built off the shore of northern Portugal in 2008. Using three Pelamis wave energy converters, the facility had about 2.3 MW of installed power capacity. During the 1970s and 1980s, Japan became a world leader in the development of ocean thermal energy conversion and remains at the forefront of this renewable energy technology.

Figure 6.1
The Annapolis Royal Generating Station in Nova Scotia, Canada is the only tidal power plant in North America. The plant has a power generation capacity of 20 MW.

(Nova Scotia Power Plant)

6.2 THE MARINE ENERGY SOURCES

Tidal energy is the only renewable energy source that is derived from the earth-moon-sun orbital system. Solar, wind, hydro, and biomass energy are directly or indirectly derived from the sun, and geothermal energy is derived from the earth. Tides are the rise and fall of sea levels due primarily to the gravitational force exerted on the earth by the moon. The gravitational force exerted on the earth by the sun and the earth's rotation are secondary causes of tides. Frequency and amplitude of tides at a given location are a function of the moon-sun alignment, the shape of the coastline and ocean floor, and the behavior of deep ocean currents. A semidiurnal tide occurs when a shoreline experiences two low tides and two nearly equal high tides each day. When a shoreline experiences one low tide and one high tide each day, the tide is called a *diurnal tide*. When a shoreline experiences two uneven tides each day, the tide is called a *mixed tide*.

Ocean currents are a direct consequence of tides, which cause the entire ocean to flow. Ocean currents are also caused by temperature and salinity gradients in ocean waters. Coriolis forces, induced by the rotation of the earth, are also responsible for ocean currents.

Ocean waves are caused by wind passing across the surface of the water. Random pressure variations in the wind near the ocean's surface transfer normal and tangential forces to the water that induce surface waves. This phenomenon is also observed on the surface of smaller bodies of water, such as lakes, ponds, and even puddles.

The oceans of the world are not isothermal. Water at and near the surface receive thermal energy from the sun that maintains shallow ocean water at temperatures higher than those at greater depths. Temperature differences in the world's oceans are typically small, ranging from about 10°C to 20°C. The temperature difference between surface and deep water is greatest in the tropical regions, but even in the tropics it is only 20°C to 25°C.

6.3 TIDAL ENERGY SYSTEMS

The conversion of tidal energy to electrical energy typically involves a **tidal barrage**, a damlike structure used to capture the energy of water that flows into and out of an *estuary* or *basin*, a region where a river flows into an ocean. A tidal barrage, illustrated in Figure 6.3, exploits the potential energy in the water's height difference, referred to as *tidal range*, between high and low tides. As the ocean level rises toward its highest point at high tide, the sluice gates are opened to allow the basin to fill. When the ocean reaches high tide, the gates are closed entrapping the water inside the basin. The basin will continue to fill as it is fed by the inland river. When the ocean reaches low tide, the tidal range is maximum, and the gates are opened allowing the stored water to rapidly flow through the turbines back to the ocean. Mechanical energy generated by the turbines is converted to electrical energy in a generator. The turbines continue to operate until the water levels in the basin and ocean are equal, at which point the tidal range is zero. The gates are then closed, and the cycle starts again. A road may be incorporated into the barrage, as depicted in Figure 6.3.

The tidal barrage just described is a one-way scheme that generates power during ebb tide (when water flows from the basin into the ocean). A tidal barrage may be designed to operate during flood tide (when water flows from the ocean to the basin), but this scheme is less efficient because the flow rate in the turbine is lower because of a slowly filling basin. The barrage may be bidirectional, generating power during the entire tidal cycle. Bidirectional systems are less efficient than one-way systems because the tidal ranges are smaller and turbines that are designed to operate in both directions are less efficient than unidirectional units.

To do a basic energy analysis of a tidal barrage, we begin with a principle of physics, which says that the gravitational potential energy of a body of mass, m, is given by the relation

$$PE = mgz \qquad\qquad (6.1)$$

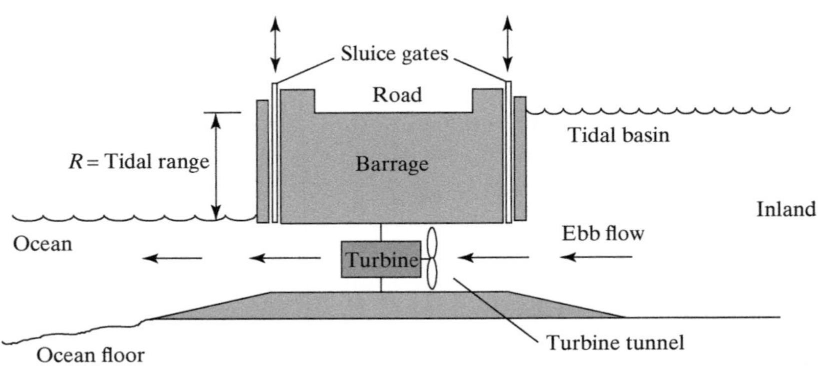

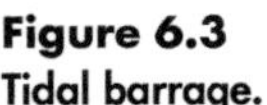

Figure 6.3
Tidal barrage.

where g is gravitational acceleration and z is vertical distance from a reference point or origin to the body's center of mass. The quantity, z, for a tidal basin is half the tidal range, $R/2$; and the mass of the water in the basin is the product of water volume, V, and water density, ρ. Thus, $m = V\rho$. Assuming a simple rectangular-shaped basin, the volume of water is the product of tidal range, R, and surface area, A, of the basin. Thus, $V = AR$. Substituting these formulas into Equation (6.1), we obtain

$$PE = \frac{1}{2} g \rho A R^2 \tag{6.2}$$

Hence, the electrical energy generated by the turbine of a tidal barrage is

$$E_{\text{elect}} = \frac{1}{2} \eta g \rho A R^2 \tag{6.3}$$

where η is the combined efficiency of the turbine, gear box, and electrical generator. Finally, the electrical output power generated by the tidal barrage is the electrical energy divided by the tidal period, T, expressed in seconds

$$P_{\text{elect}} = \frac{\eta g \rho A R^2}{2T} \tag{6.4}$$

EXAMPLE 6.1

The tidal range of a tidal barrage is 4.6 m, and the surface area of the basin is 120 acre. If the tidal barrage has eight turbines, find the electrical output power for a tidal period of 10 h. The combined efficiency of the barrage is 0.30. For the density of ocean water, use $\rho = 1025 \ \text{kg/m}^3$.

SOLUTION

First, we convert the surface area of the basin, A, to units of m^2,

$$A = 120 \ \text{acre} \times 4046.95 \ \text{m}^2/\text{acre} = 4.86 \times 10^5 \ \text{m}^2$$

The tidal period in units of s is,

$$T = 10 \ \text{h} \times 3600 \ \text{s/h} = 3.60 \times 10^4 \ \text{s}$$

Using Equation (6.4), the electrical output power of one turbine is

$$P_{\text{elect}} = \frac{\eta g \rho A R^2}{2T}$$

$$= \frac{(0.30)(9.81 \ \text{m/s}^2)(1025 \ \text{kg/m}^3)(4.86 \times 10^5 \text{m}^2)(4.6 \ \text{m})^2}{2(3.60 \times 10^4 \text{s})}$$

$$= 4.31 \times 10^5 \ \text{W} = 431 \ \text{kW}$$

Thus, the total electrical output power of the tidal barrage is

$$P_{\text{elect,total}} = 8(4.31 \times 10^5 \text{W})$$

$$= 3.45 \times 10^6 \ \text{W} = 3.45 \ \text{MW}$$

6.4 OCEAN ENERGY SYSTEMS

In this section, we consider the fundamental engineering principles of each type of ocean energy system—ocean currents, ocean waves, and ocean thermal energy.

6.4.1 Ocean Currents

As stated in Section 6.2, ocean currents are a consequence of tides, which cause the entire ocean to flow. A system that utilizes this phenomenon is called a **tidal stream generator** (TSG). In a tidal stream generator, kinetic energy of moving water is converted to mechanical energy in a turbine, which drives a generator to produce electricity. In principle, this technology is the same as wind power, but the fluid is ocean water instead of air. Accordingly, the available power in an ocean current can be found using the same relation that is used for finding the available power in wind

$$P_{\text{water}} = \frac{1}{2}\rho A v^3 \tag{6.5}$$

where ρ is density of ocean water, A is area swept out by the turbine blades, and v is water velocity. Like wind turbines, tidal stream generators are subject to the Betz limit, so the maximum theoretical power that a turbine can extract from an ocean current is

$$P_{\text{water,max}} = C_p P_{\text{water}} \tag{6.6}$$

where C_p is the power coefficient, which has a value of 16/27. The electrical power generated by a tidal stream generator is given by the relation

$$P_{\text{elect}} = \eta W_{\text{water}} = \frac{1}{2}\eta\rho A v^3 \tag{6.7}$$

where η is the combined efficiency of the turbine, gear box, and electrical generator. Note that η includes the power coefficient, C_p, which is the efficiency of the turbine alone. Equation (6.7) is the equivalent of Equation (3.20), the relation for

---

EXAMPLE 6.2

The turbine of a tidal stream generator has a blade diameter of 7.5 m. The combined efficiency of the system is 0.40. If the velocity of the ocean current is 1.8 m/s, find the electrical output power.

SOLUTION

The area swept out by the turbine blades is

$$A = \pi D^2/4$$
$$= \pi(7.5\text{ m})^2/4$$
$$= 44.2\text{ m}^2$$

Using Equation (6.7), the electrical output power is

$$P_{\text{elect}} = \tfrac{1}{2}\eta\rho A v^3$$
$$= \tfrac{1}{2}(0.40)(1025\text{ kg/m}^3)(44.2\text{ m}^2)(1.8\text{ m/s})^3$$
$$= 5.28 \times 10^4\text{ W} = 52.8\text{ kW}$$

wind turbines, for tidal stream generators. Tidal stream generators can be secured to the ocean floor by dedicated structures or integrated into existing structures such as bridges.

6.4.2 Ocean Waves

Ocean waves are caused by winds across the surface of the water. Winds are primarily caused by uneven solar heating of the earth, so this form of ocean energy is an indirect form of solar energy. To design a system for generating electrical power, engineers need to know the available power in ocean waves. Ocean waves are a complex phenomenon, but a simplified analysis that treats an ocean wave as sinusoidal, as illustrated in Figure 6.4, yields a useful approximation for the available power.

Referring to Figure 6.4, the quantities H and λ represent the wave peak-to-peak amplitude (height) and wavelength, respectively. The velocity, v, at which the wave moves across the ocean surface can be expressed in terms of wavelength as

$$v = \lambda/T \tag{6.8}$$

where T is the period of the wave, defined as the time it takes for successive peaks or troughs to pass a fixed point. Wave period, T, is the reciprocal of wave frequency, f, so $T = 1/f$. The power of ocean waves is expressed in terms of *energy flux*, defined as the average power per unit length of wave crest, expressed in units of W/m. This important quantity is the average energy per second that passes under one meter of wave crest from the ocean surface to the seabed. Energy flux decreases exponentially with depth such that nearly all of the energy of a wave is found near the surface.

For deep ocean waves, it can be shown that the energy flux, J, is given by the expression

$$J = \frac{\rho g^2}{32\pi}TH^2 \tag{6.9}$$

where ρ is seawater density and g is gravitational acceleration. For the mathematical steps leading to Equation (6.9), consult the readings at the end of this chapter. Electrical output power for one wave is the product of overall efficiency of the wave energy conversion system, η, energy flux, J, and wave crest length, L. Thus

$$P_{\text{elect}} = \eta JL \tag{6.10}$$

Figure 6.4
Sinusoidal ocean wave.

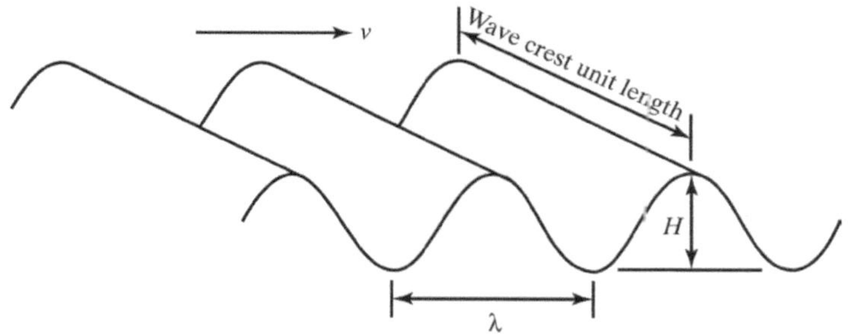

Figure 6.5
Types of wave energy
converters: (a) wave-
activated device,
(b) overtopping device,
(c) oscillating column.

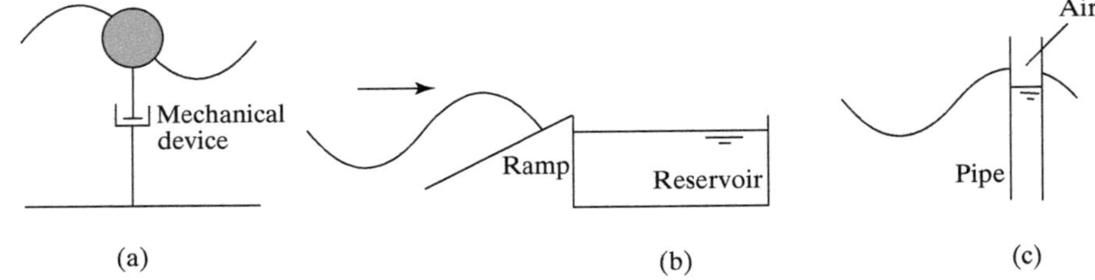

A variety of technologies can be employed to convert wave energy to electrical energy. Referred to as **wave energy converters**, these technologies can be broadly classified into three groups: *wave-activated devices*, *overtopping devices*, and *oscillating water columns*. In a wave-activated device, wave motion is directly transferred to mechanical motion of some kind. An overtopping device incorporates a ramp or tapered channel that forces water of incoming waves to spill into a reservoir. In a manner similar to hydropower plants, stored water is directed back to the ocean through a turbine. In an oscillating water column, waves cause the water surface in a partially submerged pipe to oscillate thereby pumping air in the pipe through a turbine. These systems are illustrated in Figure 6.5.

EXAMPLE 6.3

Far off shore, the wave period and height of an ocean wave are 6.0 s and 3.2 m, respectively, and the wave crest length is 80 m. Find the electrical output power for ten waves if the overall efficiency of the wave energy converter is 0.25.

SOLUTION

Using Equation (6.9), the energy flux, J, is

$$J = \frac{\rho g^2}{32\pi} T H^2$$

$$= \frac{(1025 \text{ kg/m}^3)(9.81 \text{ m/s}^2)^2(6.0 \text{ s})(3.2 \text{ m})^2}{32\pi}$$

$$= 6.03 \times 10^4 \text{ W/m}$$

Using Equation (6.10), the electrical output power of one wave is

$$P_{\text{elect}} = \eta J L$$

$$= (0.25)(6.03 \times 10^4 \text{ W/m})(80 \text{ m})$$

$$= 1.21 \times 10^6 \text{ W} = 1.21 \text{ MW}$$

Thus, the electrical output power for ten waves is

$$P_{\text{elect,total}} = 10(1.21 \text{ MW})$$

$$= 12.1 \text{ MW}$$

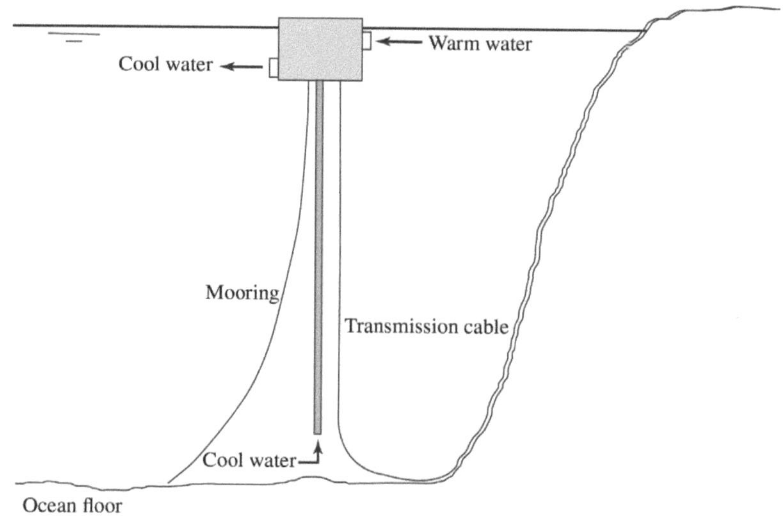

Figure 6.6
Floating ocean thermal energy conversion plant.

6.4.3 Ocean Thermal Energy

Ocean thermal energy conversion (OTEC) utilizes temperature differences between shallow and deep seawater. Water is semitransparent, so seawater transmits solar radiation to some distance below the surface. Most solar energy is absorbed by seawater within a few meters of the surface. At somewhat greater depths, seawater temperature decreases slightly due to turbulent mixing caused by winds and waves. Between depths of about 200 m and 1000 m, seawater temperature declines sharply throughout much of the ocean due to cold, dense water that sinks at the polar regions and flows to the equator. From a depth of about 1000 m to the ocean floor, seawater temperature is nearly constant at about 2°C. In deep oceans, the temperature difference between shallow and deep water can be as high as 25°C. A temperature difference in seawater can be exploited to run a heat engine. A floating OTEC plant is illustrated in Figure 6.6.

Basic heat engines and their application to geothermal power generation were presented in Chapter 5. In principle, a heat engine employed in OTEC and geothermal applications is the same. Hence, the same equations used for a basic energy analysis of a geothermal power plant can be used for analyzing an OTEC plant. (Refer to Section 5.3.) However, the temperature difference in OTEC systems is relatively small, resulting in very low thermal efficiencies, as pointed out in the following example.

EXAMPLE 6.4

An OTEC plant operates in a region where the temperatures of the surface and deep ocean waters are 30°C and 11°C, respectively. Find the maximum possible thermal efficiency of a heat engine that is employed by this system.

SOLUTION

Using Equation (5.3), the Carnot efficiency of an OTEC system in which the temperatures of the deep and shallow water, respectively, are 11°C and 30°C, is

(continued)

$$\eta_{\text{th,ideal}} = 1 - \frac{T_{\text{L}}}{T_{\text{H}}}$$

$$= 1 - \frac{(11 + 273)\text{K}}{(30 + 273)\text{K}}$$

$$= 0.063 \ (6.3 \text{ percent})$$

Accounting for energy losses, the actual thermal efficiency of a heat engine employed in an OTEC power plant typically ranges from 2 percent to 3 percent.

PRACTICE!

1. A tidal barrage has a tidal range of 3.8 m, and the surface area of the basin is 475 acre. If the tidal barrage has five turbines, find the electrical output power for a tidal period of 11 h. The combined efficiency of the barrage is 0.26.

   Answer: 4.58 MW

2. The blade diameter of the turbines of a tidal stream generator is 5.0 m. The combined efficiency of the system is 0.35. If the velocity of the ocean current is 1.3 m/s, find the total electrical output power if the system consists of six turbines.

   Answer: 46.4 kW

3. In deep ocean water, the wave period and height of an ocean wave are 5.2 s and 2.9 m, respectively, and the wave crest length is 60 m. Find the electrical output power for three waves if the overall efficiency of the system is 0.20.

   Answer: 1.54 MW

4. Consider an OTEC plant with an ideal thermal efficiency of 0.070. If the temperature of deep ocean water is 8°C, what is the temperature of ocean water at the surface?

   Answer: 29°C

PROFESSIONAL SUCCESS—TIDAL POWER GENERATION IN THE NEW YORK HARBOR

The East Channel of the East River is a tidal strait connecting Long Island Sound with the Atlantic Ocean in New York Harbor. Strong water currents that are generated in the strait change direction between flood tide and ebb tide about four times each day, making the strait a good location for a tidal power plant.

In 2012, the Federal Energy Regulatory Commission issued a pilot project license to Verdant Power, an energy company based in New York, to build a tidal power plant in the East River near Roosevelt Island. Known as the Roosevelt Island Tidal Energy (RITE) project, this is the first tidal power plant to be commercially licensed in the United States. The duration of the pilot license is ten years.

When the RITE power plant is fully built out by 2015, it will consist of 30 turbines that generate a total of 1050 kW of electrical power for approximately 9500 New York residents. The turbines are designed with passive yawing so that they generate power during both flood and ebb tides. When the water velocity exceeds approximately 1.0 m/s, the turbine blades begin to rotate. When the tide changes direction after about six hours, the turbines automatically yaw to take advantage of the tidal current flowing in the opposite direction. This cycle is repeated about four times each day. The system not only generates electrical power, but also collects useful data on environmental impacts on fish and river sedimentation.

6.5 ENVIRONMENTAL CONSIDERATIONS

Like the other renewable energy forms discussed in this book, marine energy systems do not emit greenhouse gases into the environment but are not without some environmental challenges. The construction of a tidal barrage across a river alters the sedimentation process and salinity level in the estuary thereby affecting marine and fowl life. Furthermore, fish may pass through the sluice gates and even through the turbines, causing injury or death. Many species of birds live on the mud flats and salt marshes of estuaries. A tidal barrage may reduce or eliminate the mud flats and salt marshes thereby displacing the fowl that live there. Construction of a barrage across an estuary impedes shipping, but the barrage design typically includes locks.

The turbine blades of tidal stream generators pose a threat to fish and other marine animals that attempt to swim through them. Emission of acoustic waves may adversely affect marine life, such as dolphins and whales, that use echolocation for communication and navigation.

The environmental impacts of wave energy systems are relatively benign. Pollution from lubricants would occur if the systems are not well sealed. Floating systems, if located in shipping lanes, present a danger to ships. Marine life is not affected, and noise generation is minimal.

Ocean thermal energy conversion systems take in cold seawater at depths below about 1000 m and take in warm seawater at depths about 20 m. To prevent altering the temperature of the warm seawater source, the mixed seawater is typically returned at depths of approximately 50 m or 60 m. Thermally altered seawater at these depths has the potential of affecting marine life. Over the lifetime of an OTEC plant, the heat exchanger will experience biofouling, the accumulation of biological organisms on the heat transfer surfaces that reduces thermal efficiency. Poisonous chemicals used to treat biofouling could be released into the ocean, affecting marine life.

SUMMARY

Marine energy is a broad term used to describe energy derived from oceans. There are two main types of marine energy: tidal energy and ocean energy. Ocean energy can be broken down further into ocean currents, ocean waves, and ocean thermal energy.

Mechanical energy from tidal motion dates back hundreds of years, but tidal and ocean energy have only recently been harnessed to generate electrical energy. The global energy-generation capacity of ocean currents is estimated to be 2000 TWh, and the global energy capacity of ocean waves might be as high as 10 TWh.

Tides are the rise and fall of sea levels due primarily to the gravitational force exerted on the earth by the moon. Ocean currents are a direct consequence of tides, which cause the entire ocean to flow. Ocean waves are caused by wind passing across the surface of the water. Ocean thermal energy is generated by a heat engine, which is driven by temperature differences between deep and shallow ocean water.

The conversion of tidal energy to electrical energy typically involves a tidal barrage, a damlike structure used to capture the energy of water that flows into and out of an estuary. Electrical power generated by a tidal barrage is primarily a function of the tidal basin depth and surface area. The available power in ocean currents is calculated using the same equation used for finding available power in wind, but the fluid is water instead of air. Electrical power generated by ocean waves is primarily a function of wave height and is also a function of wave period. Heat engines that generate energy due to temperature differences in ocean water are limited by the Carnot efficiency. Due to small temperature differences typically found in the oceans, ocean thermal energy conversion systems have very low thermal efficiencies.

Tidal barrages affect marine and fowl life in and around estuaries. Turbine blades of a tidal stream generator have the potential of injuring or killing fish that attempt to pass through the blades. Wave energy systems, if not well sealed, might release lubricants into the ocean. Seawater that is thermally altered by ocean thermal energy conversion systems is potentially harmful to nearby marine life.

KEY TERMS

marine energy	ocean thermal energy	tidal energy
ocean currents	conversion	tidal stream generator
ocean energy	ocean waves	wave energy converters
ocean thermal energy	tidal barrage	

SUGGESTED READING

ABDULLAH, M.O., *Applied Energy—An Introduction*, Boca Raton, FL: CRC Press, 2013.

BOYLE, G., Ed., *Renewable Energy—Power for a Sustainable Future* 3rd Ed, Oxford: Oxford University Press, 2012.

BREEZE, P., *Power Generation Technologies*, Burlington, MA: Elsevier, 2005.

HAGEN, K.D., *Introduction to Engineering Analysis* 4th Ed, Upper Saddle River, NJ: Prentice Hall, 2014.

HERBICH, J.B., *Handbook of Coastal Engineering*, New York, NY: McGraw-Hill, 2000.

KHARTCHENKO, N.V. and V.M. KHARTCHENKO, *Advanced Energy Systems* 2nd Ed., Boca Raton, FL: CRC Press, 2014.

KREITH, F., *Principles of Sustainable Energy Systems* 2nd Ed., Boca Raton, FL: CRC Press, 2014.

McCORMICK, M.E., *Ocean Wave Energy Conversion*, Mineola, NY: Dover Publications, 2007.

WATERS, R., "Energy from Ocean Waves—Full Scale Experimental Verification of a Wave Energy Converter," PhD Dissertation, Uppsala Universitet, Uppsala, Sweden, 2008.

ZOOBA, A.F. and R.C. BANSAL, Eds., *Handbook of Renewable Energy Technology*, Hackensack, NJ: World Scientific, 2011.

PROBLEMS

For the following problems, it is recommended that you use the general analysis procedure of (1) problem statement, (2) diagram, (3) assumptions, (4) governing equations, (5) calculations, (6) solution check, and (7) discussion. This procedure is covered in Chapter 3 of Hagen, K.D., *Introduction to Engineering Analysis* 4th Ed.

Tidal Energy

6.1 The tidal range of a tidal barrage is 4.8 m, and the surface area of the basin is 450 acre. If the tidal barrage has 12 turbines, find the electrical output power for a tidal period of 10 h. For the combined efficiency of the barrage, let $\eta = 0.27$.

6.2 The basin surface area and tidal range of a tidal barrage are 375 acre and 5.1 m, respectively. The barrage has eight turbines, each with a combined efficiency of 0.35. If the tidal period is 12 h, calculate the electrical output power of the barrage.

6.3 Consider a tidal barrage with five turbines, each with a combined efficiency of 0.40. On average, the water level in the tidal basin experiences high tide 2.6 times per day. If the surface area of the basin is 185 acre, find the electrical output power of the barrage.

6.4 A small tidal barrage with two turbines provides electrical power to a nearby town. The tidal range is 2.5 m, and the surface area of the basin is 60 acre. Find the electrical output power if the turbines have a combined efficiency of 0.22 and the tidal period is 11 h.

6.5 The required electrical output power of a tidal barrage is 45 MW. The tidal range is 4.6 m, and the tidal period is 10.4 h. If the barrage has ten turbines, each with a combined efficiency of 0.30, find the required surface area of the basin.

6.6 The surface area of a tidal basin is 275 acre. The barrage incorporates six turbines, each with a combined efficiency of 0.33. If the tidal period is 10.5 h, what is the required tidal range for an electrical output power of 5.7 MW?

Ocean Energy

6.7 The turbine of a tidal stream generator has a blade diameter of 4.8 m, and the efficiency of the system is 0.42. Find the electrical output power for an ocean current velocity of 2.1 m/s.

6.8 A certain ocean current has an average velocity of 1.9 m/s. A tidal stream generator with an efficiency of 0.55 is placed in this current. Assuming the ocean current is steady, find the electrical energy generated during a three-year period. The blade diameter of the turbine is 3.8 m.

6.9 The diameter of the turbine blades of a tidal stream generator is 4.75 m, and the efficiency of the system is 0.45. If the system consists of 18 turbines, find the electrical output power for an ocean current velocity of 2.6 m/s.

6.10 The electrical output power requirement of a tidal stream generator is 1.75 MW. Each turbine in the system has a blade diameter of 4.9 m and an efficiency of 0.42. For an ocean current velocity of 2.8 m/s, how many turbines are required?

6.11 The equation for finding available power in wind and ocean currents is identical. For the same turbine blade diameter and fluid velocity, how much more power is available in an ocean current than wind?

6.12 A wave energy converter operates far off shore where the wave period and wave height of ocean waves are 5.2 s and 3.8 m, respectively. The wave crest length is 60 m, and the overall efficiency of the wave energy converter is 0.20. Find the electrical output power for one wave.

6.13 The overall efficiency of a wave energy converter is 0.27. The system operates in ocean waters where the wave crest length is 75 m, the wave height is 2.4 m, and the wave period is 6.5 s. Find the total electrical output power for five waves.

6.14 The required electrical power capacity of a wave energy converter is 4.75 MW. The system is to operate in deep ocean waters where the wave period and wave height are 5.3 s and 4.8 m, respectively. If the overall efficiency of the system is 0.18, how many 75-m-long waves must the system use to generate the required power?

6.15 A wave energy converter in the design stage has an estimated overall efficiency of 0.22. The required electrical power capacity of the system is 7.0 MW. Find five combinations of wave height and wave period that will meet the power capacity requirement if the system is designed to exploit two waves with a wave crest length of 100 m.

6.16 An ocean thermal energy conversion system operates between seawater temperatures of 16°C and 29°C. Find the maximum possible thermal efficiency of the system.

6.17 Warm ocean water at 28°C supplies an OTEC system 500 kJ of thermal energy while 490 kJ of heat is rejected to 19°C ocean water. Find the output work of the turbine, the actual thermal efficiency, and the Carnot efficiency of the OTEC system.

6.18 For an ideal thermal efficiency of at least 0.035, find the minimum warm ocean temperature for an OTEC power plant that rejects heat to 17°C ocean water.

6.19 Do research and find the warmest ocean water ever recorded. Exclude water that was randomly heated by volcanic or geothermal effects. Based on this source temperature and a sink temperature of 15°C, what is the maximum thermal efficiency that an OTEC power plant could have?

6.20 Write an essay on the Pelamis wave energy converter. This device consists of a series of partially submerged cylindrical sections connected by hinged joints that, as the sections move relative to one another on the ocean's surface, deliver hydraulic fluid to motors that drive electrical generators.

Environmental Considerations

6.21 Write an essay on the environmental impacts of the La Rance tidal power plant in La Rance, France. Prior to 2011, this plant was the oldest tidal power barrage in the world.

6.22 The first tidal power plant in North America is the Annapolis Royal Generating Station in Nova Scotia. This plant operates on an inlet of the Bay of Fundy. Write an essay on the environmental issues associated with this facility.

6.23 Both geothermal power plants and OTEC power plants thermally alter their surroundings. Write an essay on the environmental impacts of this phenomenon.

7 Biomass

Objectives

After reading this chapter, you will have learned:
- What biomass is
- About the biomass energy source
- How the different types of biomass power plants work
- How to do a basic energy analysis of a biomass power plant
- The environmental impacts of biomass energy

7.1 INTRODUCTION

Biomass is the earth's organic matter found within the biosphere, the thin layer near the earth's surface. The organic matter may be waste from crops or wood waste from processing plants or forestry operations, or the matter may be grown specifically for use in heating applications or power generation. Another form of biomass is animal waste. Energy stored in biomass is naturally recycled through a series of chemical and physical conversion processes in the soil and surrounding atmosphere, which is the reason that energy from biomass is considered renewable. A part of this energy can be captured by intervening at the right time when biomass is available as fuel. Of course, the origin of the energy that drives this cyclic process is the sun, and the key mechanism of the process is photosynthesis. The carbon dioxide released by the burning of fossil fuels came from carbon that was locked by photosynthesis in hydrocarbon materials millions of years ago. Burning of biomass releases as much carbon dioxide into the atmosphere as fossil fuels, but, unlike fossil fuels, biomass does not release *new carbon* into the atmosphere. Thus, according to some environmental scientists, burning of biomass makes no net contribution to atmospheric carbon dioxide. (See Section 7.4.) Biomass can be directly converted to energy by burning it to heat a living space or to heat water to run a turbine for generating electricity. Alternatively, biomass can first be converted to a secondary substance that is subsequently combusted or combined with other substances prior to combustion.

In 2010 the global power consumption has been estimated to be about 10^{13} W. The earth's annual power generation via the photosynthesis process is approximately 10^{15} W. The total photosynthetic efficiency, the ratio of solar energy absorbed to chemical energy generated, of the earth's plants is less than 2 percent.

Humankind has utilized biomass in the form of wood and animal wastes for heating and cooking for centuries, but the use of biomass for electrical power generation is a relatively new development. Because electrical power generation is the focus of the other renewable energy resources discussed in this book, electrical power generation from biomass will be emphasized here.

7.2 THE BIOMASS ENERGY SOURCE

The ultimate source of energy from biomass is the sun. Thus, like other renewable energy sources discussed in this book, energy from biomass is an indirect form of solar energy. The essential mechanism by which energy from the sun is converted to biomass energy is **photosynthesis**. Photosynthesis is the process used by plants to convert light energy to chemical energy. The essential features of photosynthesis can be represented by the simple chemical equation

$$CO_2 + H_2O + energy \rightarrow [CH_2O] + O_2 \tag{7.1}$$

From Equation (7.1) we see that plants convert carbon dioxide, water, and energy to a carbohydrate molecule and oxygen. The carbohydrate molecule represents sugars, starches, cellulose, and other matter that, upon combustion, releases the chemical energy of the molecule. It is instructive to note that this process is the reverse of combustion or decomposition. When a plant burns or decays, oxygen is consumed, carbon dioxide and water are produced, and energy in the form of heat is released. This reverse process delineates how energy is derived from biomass.

When biomass or fossil fuel undergoes combustion, the fuel's chemical energy is released as thermal energy. For a unit mass of fuel burned, a certain amount of thermal energy is released, an amount referred to as *energy content or heat content* of the fuel. Other terms that are often used are *energy density* and *specific energy*. The energy content of a variety of biomass fuels is given in Table 7.1. It is standard practice to report the energy content of fuels in terms of *lower heating value* and *higher heating value*. Lower heating value is defined as the amount of heat released by burning a specific amount of fuel at 25°C and returning the temperature of the combustion products to 150°C. Higher heating value is defined as the amount of heat released by burning a specific amount of fuel at 25°C and returning the temperature of the combustion products to the initial temperature of 25°C, which takes into account the latent heat of vaporization of water in the combustion products. As a comparison of the heating content of biomass with fossil fuels, the lower and higher heating values of natural gas are approximately 47 MJ/kg and 52 MJ/kg, respectively, and the lower and higher heating values of coal are approximately 23 MJ/kg and 24 MJ/kg, respectively.

Biomass can be divided into six categories: agricultural biomass, forest biomass, energy plantation (crops), marine biomass, animal waste, and municipal waste. Agricultural biomass refers to residues from agricultural crops such as stalks, branches, leaves, prunings, and straw. Biomass in this category also includes agricultural processing byproducts, such as residues from fruit pits, olive pits, and

Table 7.1 Energy Content of Various Biomass Fuels

Fuel type	Energy content (MJ/kg)	
	Higher heating value (HHV)	**Lower heating value (LHV)**
Agricultural residues		
Corn stalks/stover	17.6–18.5	16.8–18.1
Sugarcane	17.3–19.4	17.7–17.9
Wheat straw	16.1–18.9	15.1–17.7
Hulls, shells, prunings	15.8–20.5	
Fruit pits	20.7–23.2	
Herbaceous crops		
Miscanthus	18.1–19.6	17.8–18.1
Switchgrass	18.0–19.1	16.8–18.6
Other grasses	18.2–18.6	16.9–17.3
Bamboo	19.0–19.8	
Woody crops		
Black locust	19.5–19.9	18.5
Eucalyptus	19.0–19.6	18.0
Hybrid poplar	19.0–19.7	17.7
Willow	18.6–19.7	16.7–18.4
Forest residues		
Hard wood	18.6–20.7	17.5–20.8
Soft wood	18.6–21.1	
Urban residues		
Municipal solid waste	13.1–19.9	12.0–18.6
Refuse-derived fuel	15.5–19.9	14.3–18.6
Newspaper	19.7–22.2	18.4–20.7
Corrugated paper	17.3–18.5	
Waxed cartons	27.3	

cotton. Forest biomass consists of wood residues from logging and thinning operations and wood byproducts from sawmills and other wood-processing industries. Energy plantation or crops refer to the growing of particular species of trees and plants that can be harvested in a short period of time and subsequently used for fuel. Energy crops require careful land management and care of the plants. The fuel from energy crops can be processed into ethanol or methanol or may be used directly for boilers, wood-burning stoves, or other heat-producing systems. Because wastes alone will not be sufficient over the long term to meet electrical power demands, energy crops grown on dedicated plantations will be required to sustain the biomass power-generation industry. Marine biomass includes seaweed, algae, floating water plants, and marine animal life. Animal waste refers to waste from livestock operations, slaughterhouses, pig farms, and poultry farms. Municipal waste includes human excreta, garbage from residential and commercial sources, and liquid sewage and effluent.

EXAMPLE 7.1

How much thermal energy is released by the complete burning of 150 kg of sugarcane? Include the latent heat of vaporization of water, and assume that at the end of the combustion process, the sugarcane combustion products are returned to a temperature of 25°C.

SOLUTION

The energy released is the product of the energy content of sugarcane and the mass burned,

$$E = HHVm$$
$$= (19.4 \text{ MJ/kg})(150 \text{ kg})$$
$$= 2910 \text{ MJ}$$

The value of *HHV* used in this example was somewhat arbitrary. As indicated in Table 7.1, we could have used any value from 17.3 MJ/kg to 19.4 MJ/kg.

7.3 BIOMASS POWER-GENERATION SYSTEMS

A variety of technologies and processes exist for converting the energy in biomass to useful energy forms. Through various physical transformations, biomass can be converted into fuels for transportation systems, or the heat released from the combustion of biomass can be used for space heating, industrial heating processes, and related applications. Because this book focuses on renewable energy for electrical power generation, technologies for using biomass to produce electricity are emphasized here.

The primary types of biomass that are used for electrical power generation are energy crops and wastes. The most common technology for converting thermal energy of burning biomass to electrical energy is a **direct-fired power plant**, illustrated in Figure 7.1. This type of power plant is similar to a conventional power plant in which coal, oil, or natural gas is burned. Burning biomass transfers heat to a boiler, and a pump circulates water in a closed loop through the boiler, turbine, and condenser. Thermal energy of high-pressure, high-temperature steam is converted to mechanical energy in the turbine, which is subsequently converted to electrical energy in the generator.

Figure 7.1
A direct-fired biomass power plant burns pure biomass.

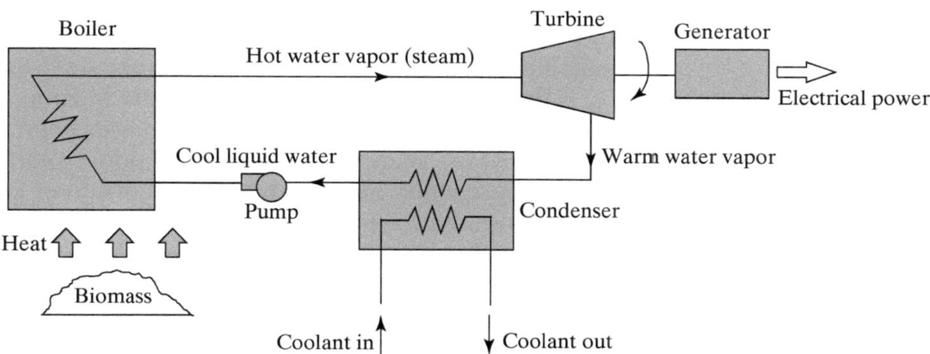

Figure 7.2
A co-fired power plant burns a mixture of coal and biomass.

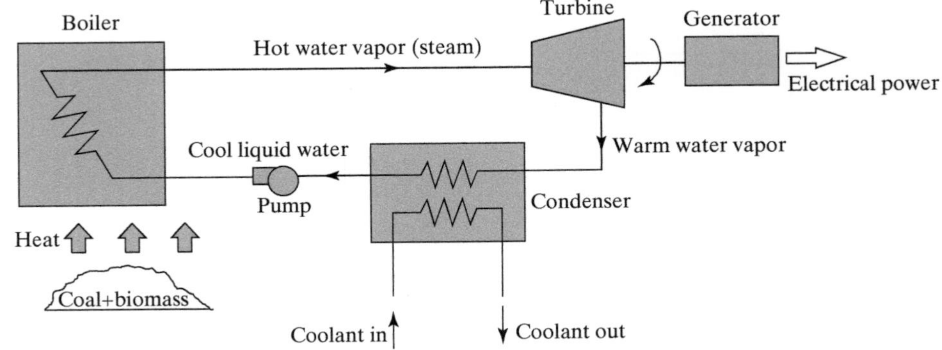

A second technology for converting thermal energy of burning biomass to electrical energy is a **co-fired power plant**. In a co-fired power plant, illustrated in Figure 7.2, a small percentage (less than 2 percent) of biomass is mixed with coal. An advantage of this technology is that it can be implemented without making modifications to the coal-fired power plant. The overall efficiency of direct-fired power plants is typically below 25 percent, but because coal burns at a higher temperature than biomass, co-fired power plants have a higher efficiency, typically around 35 to 40 percent.

The fundamental engineering principles on which direct-fired and co-fired power plants operate are identical to those of conventional power plants. Note that the only difference in Figures 7.1 and 7.2 is the fuel source. The direct-fired plant burns biomass only, whereas the co-fired plant burns a mixture of coal and biomass. The other basic components in the power plants are identical.

A thorough energy analysis of a direct-fired or co-fired power plant, or a conventional power plant for that matter, requires a knowledge of thermodynamics and other engineering sciences that are beyond the scope of this book. For our purposes, a simplified but meaningful energy analysis is presented that pertains not only to a biomass power plant but a conventional power plant as well. The following material on basic energy analysis of biomass power plants closely parallels the corresponding material in Chapter 5 on geothermal power plants.

A **heat engine** is a device that converts heat to work, and this is precisely what a power plant does. This definition applies to both biomass power plants and conventional power plants. A power plant converts heat to shaft work of a turbine, which is converted to electrical energy. As illustrated in Figure 7.3, a heat engine receives an amount of heat, Q_{in}, from a high-temperature source and converts a portion of that heat to work, W_{out}. The heat engine rejects the remaining heat, Q_{out}, to a low-temperature sink. In a conventional power plant, Q_{in} is the heat supplied to the boiler from burning a fossil fuel, such as coal, oil, or natural gas. In the case of a nuclear power plant, heat is supplied by the decay of a radioactive material. In a biomass power plant, Q_{in} is the heat supplied by the combustion of biomass. In all types of power plants, Q_{out} is heat rejected to the environment. Typically, Q_{out} is heat rejected to the atmosphere via a cooling tower, but Q_{out} can also be heat rejected to a natural body of water, such as an ocean, lake, or river via a heat exchanger. The quantity W_{out} is the shaft work (energy) generated by the turbine.

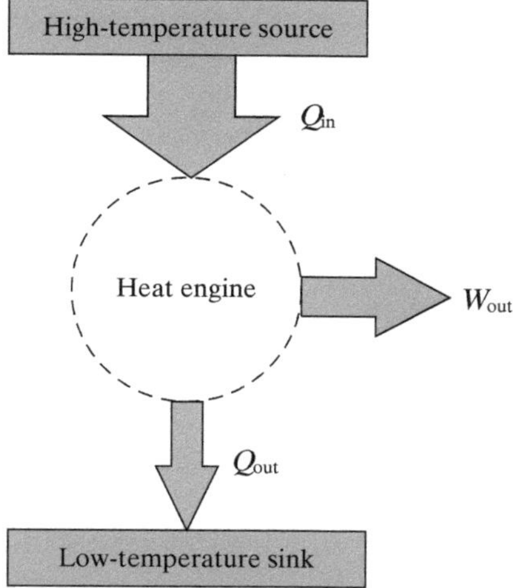

Figure 7.3
A heat engine converts a portion of the heat it receives from a high-temperature source to work and rejects the remaining heat to a low-temperature sink.

The **first law of thermodynamics** states that energy is conserved. Thus, for a heat engine, the heat supplied must equal the output work plus the rejected heat. Expressed mathematically, the first law for a heat engine is written as

$$Q_{in} = Q_{out} + W_{out} \qquad (7.2)$$

where each quantity is expressed in units of the joule (J). Stated in terms of power, these quantities can also be expressed in units of the watt (W). **Thermal efficiency** of a heat engine, denoted by η_{th}, is defined as the shaft work of the turbine divided by the heat input,

$$\eta_{th} = \frac{W_{out}}{Q_{in}} \qquad (7.3)$$

The **second law of thermodynamics** states that is impossible for a heat engine to convert all the heat it receives from a high-temperature source to work. Hence, the value of η_{th} is limited to values less than 1. A well-known theorem from thermodynamics states that for an *ideal* heat engine operating between source and sink temperatures of T_H and T_L, respectively, the maximum thermal efficiency is given by the relation

$$\eta_{th,ideal} = 1 - \frac{T_L}{T_H} \qquad (7.4)$$

where T_H and T_L are expressed in absolute temperature units of kelvin (K) or rankine (R). The thermal efficiency given by Equation (7.4) is the maximum possible thermal efficiency a heat engine can have and is referred to as the **Carnot efficiency**, in honor of the French engineer Sadi Carnot.

Turbine shaft work is converted to electrical energy by the generator, which means that the generator also has an efficiency. **Generator efficiency** is defined as electrical energy divided by shaft work,

$$\eta_{gen} = \frac{E_{elect}}{W_{out}} \qquad (7.5)$$

Generator efficiencies typically exceed 0.90. The quantities on the right-hand side of Equation (7.5) can also be expressed in terms of power.

In addition to thermal efficiency and generator efficiency, there is also an efficiency associated with the combustion of the fuel. **Combustion efficiency, η_{comb},** is defined as energy released during combustion divided by the heating value of the fuel times the mass, m, of the fuel,

$$\eta_{comb} = \frac{Q_{in}}{HVm} \tag{7.6}$$

The quantity Q_{in}, the same quantity that appears in Equation (7.2) and (7.3), is the actual energy released during combustion, and HV is the heating value of the fuel as given in Table 7.1. Equation (7.6) can also be expressed in rate form by replacing Q_{in} with $\dot{Q}_{in}$ and m with $\dot{m}$, the rate at which the biomass fuel is burned. The product of combustion, thermal, and generator efficiencies defines the **overall efficiency, $\eta_{overall}$.** Noting the cancellation of the quantities Q_{in} and W_{out}, we have

$$\eta_{overall} = \eta_{comb}\eta_{th}\eta_{gen} = \frac{Q_{in}}{HVm}\frac{W_{out}}{Q_{in}}\frac{E_{elect}}{W_{out}} = \frac{E_{elect}}{HVm} \tag{7.7}$$

The higher heating value, HHV, is typically used for HV in Equation (7.7). The relation for overall efficiency can also be expressed in terms of power by replacing E_{elect} with P_{elect} and replacing m with $\dot{m}$,

$$\eta_{overall} = \frac{P_{elect}}{HV\dot{m}} \tag{7.8}$$

where P_{elect} is expressed in the power unit of W and $\dot{m}$ is expressed in units of kg/s.

The use of Equation (7.8) enables us to calculate the *instantaneous* electrical output power of a biomass power plant, that is, the electrical output power of a plant at a moment of time for a given set of conditions. Of particular interest to engineers who design and analyze biomass power plants is the *total amount of electrical energy* that a biomass plant generates during a specific time period.

The plant **capacity factor,** CF, introduced in Chapter 1, is defined as the ratio of the actual electrical energy generated by a power plant during a specific time period to the amount of electrical energy generated by the plant during the same time period assuming the plant operates continuously at the rated capacity. Since power is defined as energy divided by time, the electrical energy generation of the plant may be expressed as

$$E_{elect} = CFP_{elect}\Delta t \tag{7.9}$$

where P_{elect} is electrical output power and Δt is the time period. Biomass power plant capacity factors vary widely, ranging from about 0.25 to 0.75.

It is important to note that plant capacity factor can be used for calculating the electrical energy generation of any power plant, including a conventional fossil-fuel plant, a nuclear plant, or a renewable energy plant. The range of values of CF depends on the type of power plant.

The following examples illustrate the use of the foregoing equations.

EXAMPLE 7.2

A direct-fired power plant burns corn stalks as fuel at a rate of 4.5 kg/s. The combustion, thermal, and generator efficiencies are 0.85, 0.30, and 0.92, respectively. Find the electrical output power of the plant. If the corn stalks burn at 475°C and the temperature of the environment is 20°C, what is the maximum possible thermal efficiency of the plant?

SOLUTION

Using Equation (7.7), the overall efficiency of the power plant is

$$\eta_{\text{overall}} = \eta_{\text{comb}}\eta_{\text{th}}\eta_{\text{gen}}$$

$$= (0.85)(0.30)(0.92)$$

$$= 0.235$$

From Equation (7.8), the electrical output power is

$$P_{\text{elect}} = \eta_{\text{overall}}HV\dot{m}_{\text{fuel}}$$

$$= (0.235)(18.5 \times 10^6\,\text{J/kg})(4.5\,\text{kg/s})$$

$$= 1.96 \times 10^7\,\text{W} = 19.6\,\text{MW}$$

Note that we used a higher heating value, *HHV*, of corn stalks from Table 7.1. Using Equation (7.4), the ideal or Carnot thermal efficiency of the power plant is

$$\eta_{\text{th,ideal}} = 1 - \frac{T_{\text{L}}}{T_{\text{H}}}$$

$$= 1 - \frac{(20 + 273)\text{K}}{(475 + 273)\text{K}}$$

$$= 0.608$$

The actual thermal efficiency of the plant is lower than the Carnot efficiency, as required by the second law of thermodynamics.

EXAMPLE 7.3

In a direct-fired power plant, switchgrass is burned at a rate of 12 kg/s at a temperature of 450°C. The combustion efficiency is 0.80, and the output power of the turbine shaft is 80 MW. If the efficiency of the electrical generator is 0.94, find the thermal efficiency, overall efficiency, electrical output power, rate of heat rejection to a 20°C atmosphere, and Carnot efficiency. If the plant capacity factor is 0.40, how much electrical energy does the plant generate in one year?

SOLUTION

The only quantity that we can calculate initially is $\dot{Q}_{\text{in}}$, the rate of heat input to the boiler from the burning biomass. Using the rate version of Equation (7.6), we have

$$\dot{Q}_{\text{in}} = HV\eta_{\text{comb}}\dot{m}$$

$$= (19.1 \times 10^6\,\text{J/kg})(0.80)(12\,\text{kg/s})$$

$$= 1.83 \times 10^8\,\text{W}$$

Knowing $\dot{Q}_{in}$, we can calculate thermal efficiency, η_{th},

$$\eta_{th} = \frac{\dot{W}_{out}}{\dot{Q}_{in}}$$

$$= \frac{80 \times 10^6 \,\text{W}}{1.83 \times 10^8 \,\text{W}}$$

$$= 0.437$$

Next, using Equation (7.7) we find overall efficiency, $\eta_{overall}$,

$$\eta_{overall} = \eta_{comb}\eta_{th}\eta_{gen}$$

$$= (0.80)(0.437)(0.94)$$

$$= 0.329$$

Using Equation (7.8), the electrical output power is

$$P_{elect} = \eta_{overall}HV\dot{m}$$

$$= (0.329)(19.1 \times 10^6 \,\text{kg/s})(12 \,\text{kg/s})$$

$$= 7.54 \times 10^7 \,\text{W} = 75.4 \,\text{MW}$$

where the value for HV is from Table 7.1. Solving for $\dot{Q}_{out}$ in Equation (7.2), we have

$$\dot{Q}_{out} = \dot{Q}_{in} - \dot{W}_{out}$$

$$= 1.83 \times 10^8 \,\text{W} - 80 \times 10^6 \,\text{W}$$

$$= 1.03 \times 10^8 \,\text{W} = 103 \,\text{MW}$$

Using Equation (7.4), the Carnot efficiency is

$$\eta_{th,ideal} = 1 - \frac{T_L}{T_H}$$

$$= 1 - \frac{(20 + 273)\text{K}}{(450 + 273)\text{K}}$$

$$= 0.595$$

The time period, Δt, in Equation (7.9) is the number of seconds in a year

$$\Delta t = 24 \,\text{h/day} \times 3600 \,\text{s/h} \times 365 \,\text{day} = 3.15 \times 10^7 \,\text{s}$$

Thus, the electrical energy generated by the plant in one year is

$$E_{elect} = CFP_{elect}\Delta t$$

$$= (0.40)(7.54 \times 10^7 \,\text{J/s})(3.15 \times 10^7 \,\text{s})$$

$$= 9.50 \times 10^{14} \,\text{J}$$

Our answer, expressed in the more commonly used electrical energy unit of kWh, is

$$E_{elect} = 9.50 \times 10^{14} \,\text{J} \times (1 \,\text{kWh}/3.6 \times 10^6 \,\text{J})$$

$$= 2.64 \times 10^8 \,\text{kWh}$$

PRACTICE!

1. How much thermal energy is released by the complete burning of 600 kg of municipal solid waste? Include the latent heat of vaporization of water, and assume that at the end of the combustion process, the combustion products of the waste are returned to a temperature of 25°C.

   Answer: 11.9 GJ

2. A co-fired power plant burns a mixture of coal and a small amount of biomass at a rate of 18 kg/s. The combustion, thermal, and generator efficiencies are 0.90, 0.35, and 0.92, respectively. Find the electrical output power of the plant. If the coal-biomass mixture burns at 850°C and the temperature of the environment is 15°C, what is the maximum possible thermal efficiency of the plant?

   Answer: 125 MW, 0.744

7.4 ENVIRONMENTAL CONSIDERATIONS

There are primarily three environmental considerations of biomass: atmospheric emissions, land use, and water use.

Unlike the other renewable energy sources discussed in this book, biomass can be considered a direct replacement for fossil fuels, particularly coal, in the respect that one combustible fuel is replaced by another. Like the burning of fossil fuels, the burning of biomass releases carbon dioxide into the atmosphere, but this carbon dioxide is taken in by the replacement biomass by photosynthesis. Thus, no net carbon dioxide is produced. But is this actually the case? A power production technology is considered to be *carbon neutral* if it produces no net increase in greenhouse gas (mainly carbon dioxide) emissions on a life cycle basis. (See Professional Success section.) The premise that biomass technologies are carbon neutral has come under scrutiny as this technology is more closely examined. Whether biomass is carbon neutral depends on the type of feedstock used, the type of power production technology employed, and the time frame specified. The carbon neutrality issue continues to be a point of controversy as biomass technologies are developed.

The combustion of biomass releases other substances besides carbon dioxide into the environment, namely, organic compounds, carbon monoxide, nitrogen oxides, and particulates. The emission of these substances is typically controlled by removing them from the power plant's flue gas. Because biomass contains no sulfur, biomass power plants do not emit sulfur dioxide as do coal-fired power plants. Unlike coal, biomass contains no toxic metals. Furthermore, the combustion of biomass produces less ash than the combustion of coal.

The environmental impacts of land use by biomass power plants depend primarily on whether the biomass is from waste or energy crops grown specifically for electrical power generation. Biomass wastes are by-products of other processes, such as logging and farming, that would have been generated anyway, so there is no significant increase in land use. However, waste products from forest and agricultural operations, if not collected properly, can result in land degradation. In farming operations it is important that sufficient crop residues are left on the land to improve soil carbon storage, maintain soil nutrient levels, and control erosion. In forest operations it is important that proper forest management practices be followed to keep the forest healthy and sustain wildlife habitat.

Energy crops and food crops share many of the same environmental challenges—watering and fertilization, crop rotation, and pest management. Many energy crops require less fertilizer and pesticides than food crops. Furthermore, perennial grasses do not require annual tilling and planting. Planting of energy crops and food crops can even be alternated to help stabilize the soil.

The cooling systems of biomass power plants and coal-fired power plants require about the same amount of water. Depending on the type of cooling system employed, the cooling water may be withdrawn from local sources, circulated through the power plant, and then discharged back into the environment. Alternatively, the cooling water may be recirculated through the plant prior to discharging it.

Obviously, water is required to produce all biomass. However, forest, agricultural, and urban wastes require no additional water, but energy crops can be water intensive depending on the type of crop grown, soil quality, and temperature. The cultivation of energy crops can adversely affect water quality through soil tillage and nutrient runoff.

PROFESSIONAL SUCCESS—LIFE CYCLE ANALYSIS AND BIOMASS POWER

Life cycle analysis (LCA), also referred to as life cycle assessment, addresses the environmental aspects and potential environmental impacts throughout a product's life cycle from raw material acquisition through production, utilization, end-of-life treatment, and disposal. This type of analysis is often referred to as a cradle-to-grave assessment. A familiar example of an application of LCA is the assessment of whether we should use paper or plastic grocery sacks.

The primary components of life cycle analysis are the compilation of relevant energy and material inputs and environmental releases, evaluation of the potential environmental impacts associated with those inputs and releases, and the interpretation of the LCA results for the purpose of making more informed decisions. In principle, all decisions that influence the environmental impact of a product or process should be evaluated using a life cycle analysis.

A life cycle analysis of a fossil-fuel power plant regards the coal or natural gas as fuel burned and therefore as energy permanently consumed. A life cycle analysis of a biomass power plant, however, does not regard the biomass as energy permanently consumed because a major assumption of the analysis is that during the lifetime of the power plant the biomass is replaced. In other words, a life cycle analysis of biomass power plants excludes fuel consumption. To make a level life cycle analysis comparison of fossil fuels and biomass, the consumption of fossil fuel must be excluded as well. A level comparison shows that a direct-fired biomass power plant consumes approximately 125 kJ/kWh of electrical energy generated, whereas a coal-fired plant consumes approximately 700 kJ/kWh of electrical energy generated, and a natural-gas fired power plant consumes approximately 1720 kJ/kWh of electrical energy generated. The primary reason for the differences is that mining and transporting fossil fuels is more expensive than harvesting and transporting biomass. Even though the thermal efficiency of fossil-fuel power plants is higher than the efficiency of biomass power plants, a life cycle analysis shows that biomass power plants are, in the broader energy sense, more efficient than fossil-fuel power plants.

SUMMARY

Biomass is the earth's organic matter found within the biosphere. This matter may be wastes from crops or wood waste from processing plants or forestry operations, or the matter may be grown specifically for heating or power generation. The sun is the ultimate energy source for biomass, and the key mechanism that drives the cycle by which biomass is produced is photosynthesis. Unlike fossil fuels, biomass is considered by some environmental scientists to be carbon neutral.

When biomass undergoes combustion, the fuel's chemical energy is released as thermal energy. The amount of thermal energy released per unit mass of fuel is the energy content or heat content of the fuel. Energy content of fuels is expressed in terms of lower heating value and higher heating value. Higher heating value includes the heat of vaporization of water, whereas lower heating value does not.

Biomass can be divided into six categories: agricultural biomass, forest biomass, energy plantation (crops), marine biomass, animal waste, and municipal waste.

The main types of biomass that are used for electrical power generation are energy crops and wastes. The most common technology for converting thermal energy of burning biomass to electrical energy is a direct-fired power plant that burns pure biomass. Another common technology is a co-fired power plant that burns coal mixed with a small amount of biomass. The fundamental engineering principles on which these power plants operate are identical to those of conventional power plants. Hence, a basic energy analysis of a biomass power plant parallels that of a conventional power plant.

The three primary environmental considerations of biomass are atmospheric emissions, land use, and water use. Combustion of biomass releases carbon dioxide into the atmosphere, but the carbon dioxide is taken in by the replacement biomass. Biomass contains no sulfur or toxic metals. With respect to land use, energy crops and food crops share many of the same environmental challenges with respect to watering and fertilization, crop rotation, and pest management. Water is required to produce biomass and for cooling systems in biomass power plants.

KEY TERMS

biomass	first law of	second law of
capacity factor	thermodynamics	thermodynamics
Carnot efficiency	generator efficiency	thermal efficiency
co-fired power plant	heat engine	
combustion efficiency	overall efficiency	
direct-fired power	photosynthesis	
plant		

SUGGESTED READING

BOYLE, G., Ed., *Renewable Energy—Power for a Sustainable Future* 3rd Ed, Oxford: Oxford University Press, 2012.

BREEZE, P., *Power Generation Technologies*, Burlington, MA: Elsevier, 2005.

HAGEN, K.D., *Introduction to Engineering Analysis* 4th Ed, Upper Saddle River, NJ: Prentice Hall, 2014.

KHARTCHENKO, N.V. and V.M. KHARTCHENKO, *Advanced Energy Systems* 2nd Ed., Boca Raton, FL: CRC Press, 2014.

KREITH, F., *Principles of Sustainable Energy Systems* 2nd Ed., Boca Raton, FL: CRC Press, 2014.

ZOOBA, A.F. and R.C. BANSAL, Eds., *Handbook of Renewable Energy Technology*, Hackensack, NJ: World Scientific, 2011.

PROBLEMS

For the following problems, it is recommended that you use the general analysis procedure of (1) problem statement, (2) diagram, (3) assumptions, (4) governing equations, (5) calculations, (6) solution check, and (7) discussion. This procedure is covered in Chapter 3 of Hagen, K.D., *Introduction to Engineering Analysis* 4th Ed.

The Biomass Energy Source

7.1 How much thermal energy is released by the complete burning of 225 kg of Miscanthus? Include the latent heat of vaporization of water, and assume that at the end of the combustion process, the Miscanthus combustion products are returned to a temperature of 25°C.

7.2 Find the thermal energy released by the complete combustion of 350 kg of hardwood forest residue. Include the latent heat of vaporization of water, and assume that at the end of the combustion process, the hardwood combustion products are returned to a temperature of 25°C.

Biomass Power-Generation Systems

7.3 Find the heat input to the boiler of a biomass power plant that burns 500 kg of municipal solid waste. Use the higher heating value, and assume a combustion efficiency of 0.85.

7.4 If the heat input per kilogram of fuel burned to the boiler of a biomass power plant is 16.5 MJ, what is the combustion efficiency if the fuel is corrugated paper?

7.5 A biomass power plant burns soft wood forest residues at a rate of 1.6 kg/s. For a combustion efficiency of 0.80, find the rate of heat transfer to the boiler.

7.6 Find the thermal efficiency of an ideal power plant that burns biomass at a temperature of 350°C and rejects its waste heat to a 20°C environment.

7.7 If the combustion and generator efficiencies of the ideal power plant in Problem 7.6 are 0.80 and 0.92, respectively, what is the overall efficiency?

7.8 A biomass power plant burns corn stalks at a rate of 4.5 kg/s. If the overall efficiency of the plant is 0.28, find the electrical output power.

7.9 The combustion and generator efficiencies of a biomass power plant are 0.88 and 0.94, respectively, and the turbine generates 20 MW of mechanical power. If the plant burns wheat straw at a rate of 6.0 kg/s, find the electrical output power, overall efficiency, and rate of heat transfer to the surroundings.

7.10 In a biomass power plant, sugarcane is burned at a rate of 2.2 kg/s. The combustion, thermal, and generator efficiencies of the plant are 0.81, 0.60, and 0.93, respectively. Find the overall efficiency, rate of heat transfer to the boiler, electrical output power, and turbine output power.

7.11 If the Carnot efficiency of a biomass power plant is 0.55 and the temperature of the surroundings is 10°C, what must the combustion temperature of the fuel be?

7.12 In a direct-fired power plant, hardwood forest residue is burned at a rate of 9.6 kg/s at a temperature of 360°C. The combustion efficiency is 0.85, and the output power of the turbine shaft is 65 MW. If the efficiency of the electrical generator is 0.94, find the thermal efficiency, overall efficiency, electrical output power, rate of heat rejection to a 20°C atmosphere, and Carnot efficiency. If the plant capacity factor is 0.30, how much electrical energy does the plant generate in one year?

7.13 A co-fired power plant burns a mixture of coal and a small amount of municipal solid waste at a rate of 8.5 kg/s. The combustion, thermal, and generator efficiencies are 0.86, 0.40, and 0.94, respectively. Find the electrical output power of the plant. If the coal-biomass mixture burns at 800°C and the temperature of the environment is 10°C, what is the maximum possible thermal efficiency of the power plant?

7.14 Consider a biomass power plant that operates as a Carnot heat engine between the temperature limits of 10°C and 300°C. The power plant burns black locust wood as fuel. Assuming combustion and generator efficiencies of 0.90 and 0.95, find the electrical output power for a unit mass per second of fuel burned.

7.15 For the power plant in Problem 7.14, find the output power of the turbine and the rate of heat transfer to the surroundings.

7.16 An inventor proposes a biomass power plant that burns fuel at a temperature of 275°C in a 20°C environment. The inventor claims that the combustion and generator efficiencies are 0.85 and 0.90, respectively. He also claims that the overall efficiency of the power plant is 0.45. Evaluate the inventor's claims.

7.17 Hulls, shells, and prunings are burned at a rate of 2.8 kg/s at a temperature of 320°C in a direct-fired power plant. The combustion efficiency is 0.89, and the output power of the turbine shaft is 45 MW. If the efficiency of the electrical generator is 0.92, find the thermal efficiency, overall efficiency, electrical output power, rate of heat rejection to a 5°C atmosphere, and Carnot efficiency. If the plant capacity factor is 0.25, how much electrical energy does the plant generate in one year?

7.18 A power plant that burns energy crops has a power-generation capacity of 1.8 MW. The plant costs $3.0 million to install. The fixed charge rate is 7.0 percent, the annual operation and maintenance cost is assumed to be 1.0 percent of the initial cost, and the levelized replacement cost is averaged over an expected 20-year lifetime. The annual cost of energy crops is $300,000, and the capacity factor is 0.25. Find the cost of energy for the plant.

Environmental Considerations

7.19 Write an essay that addresses the environmental impacts of the release of nitrogen oxides and particulate matter by biomass power plants.

7.20 Whether biomass is carbon neutral continues to be argued. Write an essay that addresses the arguments on both sides of the debate.

7.21 Write an essay on the environmental impacts of bio fuel for the transportation industry.

A The Trapezoidal Rule

The trapezoidal rule is a numerical integration technique that involves subdividing the area under a curve into narrow trapezoidal-shaped areas. The curve (function) is approximated in each subinterval by a straight line, as shown in Figure A.1.

Figure A.1
The trapezoidal rule approximates the area under a curve.

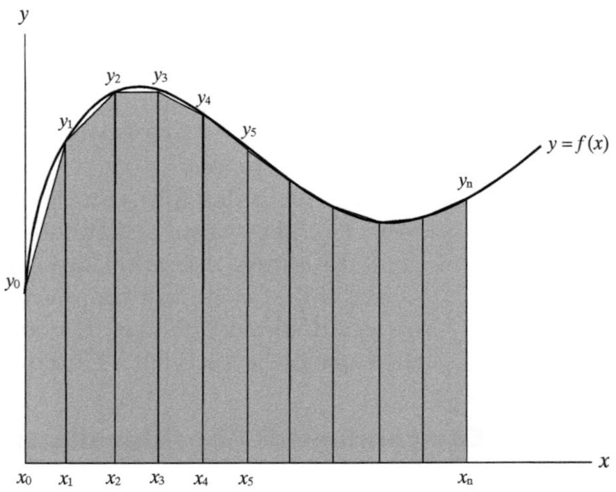

The integral of a function over the interval $x = a$ to $x = b$ is

$$I = \int_a^b f(x)\,dx$$

which is the area under the curve defined by $f(x)$ over the interval. We let $a = x_0$ and $b = x_n$ and define the width, h, of each subinterval as

$$h = (b - a)/n$$

where n is the number of subintervals. The sum of the areas of the trapezoids is

$$S = \frac{1}{2}(y_0 + y_1)h + \frac{1}{2}(y_1 + y_2)h + \ldots + \frac{1}{2}(y_{n-1} + y_n)h$$

which, after simplification, can be written as

$$S = \frac{h}{2}(y_0 + 2y_1 + 2y_2 + \ldots + 2y_{n-1} + y_n) \qquad \textbf{(A.1)}$$

The trapezoidal rule becomes more accurate as the width, h, of the subintervals decreases.

EXAMPLE

Using the trapezoidal rule, find the area under the curve defined by the points in Table A.1.

Table A.1

x	y
0	2.00
1	3.00
2	6.00
3	11.00
4	18.00

SOLUTION

There are four subintervals, so $h = (b - a)/n = (4 - 0)/4 = 1$. Using Equation (A.1), the approximate area under the curve is

$$S = (1/2)[2.00 + 2(3.00) + 2(6.00) + 2(11.00) + 18.00]$$

$$= 30.00$$

To illustrate the accuracy of the trapezoidal rule, we chose data points to precisely describe the function

$$f(x) = 2 + x^2$$

which can be integrated exactly

$$I = \int_0^4 (2 + x^2)\,dx$$

$$= 88/3 \approx 29.33$$

As shown in Figure A.2, the trapezoidal rule overpredicts the area slightly. The error is 2.27 percent.

Figure A.2
The trapezoidal rule for
$f(x) = 2 + x^2$.

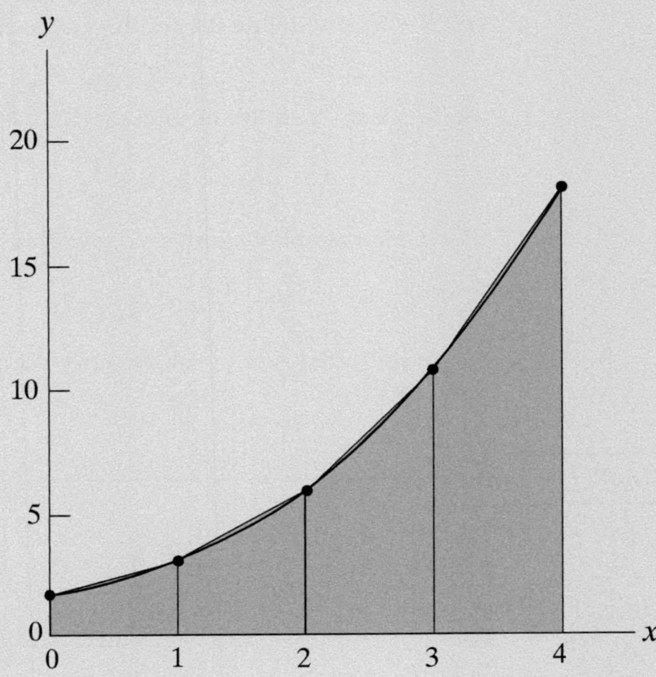

B Answers to Selected Problems

Chapter 1

1.2 244 MJ
1.4 44.0 kW, 0.0440 MW
1.6 0.627 kW
1.8 81 kW
1.10 38.8 m^2
1.12 9.83×10^6 Btu
1.14 53.4 MW
1.16 7.02 y
1.18 \$0.0476/kWh

Chapter 2

2.2 $-7.91°$, 819.6 min
2.4 786 W/m^2
2.6 958 W/m^2
2.8 16.4 kW, 21.5 MJ/m^2, 387 MJ
2.10 243 kWh
2.12 821 kWh
2.14 70.6°
2.16 1536 kWh
2.18 $\beta_{2,\text{opt}} \approx 66°$, so the angle between the roof and panel is 21°. 75.3 kWh
2.22 63.2°C
2.24 1527 W, 1.50 MW/m^2, 0.756
2.26 0.825
2.28 1175, 3.31 acre

Chapter 3

3.2 100 m
3.4 Chicago: 983 kW,
Denver: 841 kW,
Las Vegas: 942 kW,
Helena: 887 kW,
Topeka: 974 kW
3.6 127.6 m^2
3.8 1.05 MW
3.10 210 kWh
3.12 11.6 m/s
3.14 1782 kWh
3.16 38

3.18 $0.088/kWh

3.20 11.8 kW

3.22 104 W; no, the proposal is not viable because the electrical output power is too low to justify the cost of the retrofit.

Chapter 4

4.2 816

4.4 2.20 GW

4.6 2.73×10^{10} kWh

4.8 126 m

4.10 1.28×10^9 kWh

4.14 2.54 MW

4.16 2.83 m/s

Chapter 5

5.2 9.26×10^5 J/kg

5.4 0.559

5.6 0.459

5.8 0.317, 0.285, 8.55 MW, 20.5 MW

5.10 0.417, 0.392, 6.58 MW, 9.8 MW, 0.510

5.12 0.0714, 0.0679, 23.8 kW, 325 kW, 0.140

5.14 0.325, 0.296, 5.92 MW, 13.5 MW, 0.527, 4.55×10^7 kWh

Chapter 6

6.2 12.5 MW

6.4 41.4 kW

6.6 4.41 m

6.8 5.76×10^5 kWh

6.10 20

6.12 863 kW

6.14 3

6.16 0.0430

6.18 27°C

Chapter 7

7.2 7.25 GJ

7.4 0.892

7.6 0.546

7.8 23.3 MW

7.10 0.452, 34.6 MW, 19.3 MW, 20.7 MW

7.12 0.385, 0.307, 61.1 MW, 104 MW, 0.537, 1.61×10^7 kWh

7.14 8.31 MW

7.16 The calculated overall efficiency is only 0.356 based on the inventor's values of η_{comb}, η_{gen}, and a Carnot efficiency of 0.465. Thus, the inventor's claim is invalid.

7.18 $0.151/kWh

Index